White Bus S[ervices]

Berkshire's Oldest In[dependent]

Written and Published by
Paul Lacey

Featuring photographs from Tony Wright
and the White Bus Archive

The original White Bus Ford 14-seaters are seen on an otherwise traffic-free Ascot High Street of the early 1920's.

**Written, designed, typeset and published by Paul Lacey, 17 Sparrow Close
Woosehill, Wokingham Berkshire, RG41 3HT**

ISBN 978-0-9567832-2-6

Printed by Biddles Books, King's Lynn, Norfolk, PE32 1SF

Deep in the Great Park, Bedford YLQ (HRO 958V) was caught by Tony Wright leaving Norfolk Farm in May 1988.

CONTENTS

Author's Note – When this story of the *White Bus* started, the streets of Windsor echoed to the larger buses of *Thames Valley* and *London General,* plus a myriad of smaller types owned by some 20 to 30 local independents. Although there are remnants of the larger concerns still to be seen, there is now only one local independent bus operator left from those times – *White Bus Services.*
Long Live The White Bus!

Foreword

Doug with the first White Bus Bedford

A book on White Bus Services has been proposed by numerous people over the years but Tony Wright's persistence paid off when he persuaded Paul Lacey to set time aside for the project a year ago.

The idea had always appealed to me but the thought of trawling through 85 years' worth of archives was somewhat daunting. However my task of locating boxes of papers and chasing off the odd mouse or two was nothing compared to Paul's and he has done an absolutely wonderful job of blowing off the dust and sorting through a somewhat fragmented (literally sometimes) record with some genuinely surprising results for me. Not only did I find out about activities the company engaged in and about changes and proposals for routes that I'd known nothing about but I also gained new information about my family and insights into the characters of my Grandfather and parents in particular their persistence in keeping the company going in often hard times.

It is now 43 years since I took over the running of the company from my father and inevitably there have been ups and downs in all that time. There were times when it seemed to be a burden (and hard on my own family) but it has been a privilege to work with so many people over the years and to have not just served the community but to be a part of that community.

I would like to thank all those many people who have contributed memories, information and photos but in particular Paul for distilling all that information and Tony for a sequence of photos spanning decades providing a stunning visual record to complement all the documents.

I hope you will enjoy reading this record of, what by now, is a dwindling number of genuine independent small bus companies as much as I have enjoyed trawling through my own memories and a new job as a proof reader!

Doug Jeatt   Winkfield   March 2015

Acknowledgements

As always with a project of this complexity, there have been many contributors in varying ways. In general I am most indebted to Doug Jeatt for allowing me to peruse the White Bus Archive, along with newer files, from which much of this history has been brought into focus, and also Office Manager Helen Taylor for her co-operation. Philip Wallis kindly helped with service changes, as did Mike Stephens, whilst other useful facts came from Ben Argrave, Neil Beswick, Derek Bradfield, Brian Coney, Michael Clancy, Mac Head, Michael Mills, Alan Munro, Ann Tarrant, Reg Westgate and my fellow members at the Provincial Historical Research Group of the Omnibus Society in response to my various appeals. The memoirs of Hilda Nugent gave me a better appreciation of the Jeatt family early life at Winkfield, whilst former drivers Les Spong, Joe Sutcliffe and Mick Fazey and family members Christine Mauler and Terry Dwan added further interesting insights.

On the photographic front, these fall into three main categories. Firstly, there are the historical items from the White Bus Archive, along with those from Christine Mauler and Hilda Nugent, the modern prints having been produced by Tony Wright. Secondly, the extensive photography of Tony Wright has formed the basis of the more modern era, and without which this book could not have been contemplated! And thirdly, there are some other views sourced from either local photographers or those mostly no longer with us but who had the foresight to photograph the fleet. As is sometimes necessary, some views have been sourced of vehicles with other operators to fill gaps, often taken many miles away. The known contributors are John Aldridge, Mervyn Annetts, Alan Cross, John Cummings, John Gillham, Sue Hester, Joe Higham, D.A. Jones, D.W.K. Jones, Philip Kelley, Alan Lambert, Bob Mack, Roy Marshall, Phil Moth, Mal Saltmarsh, R.H.G. Simpson, Chris Spencer and Philip Wallis. Thanks are also due to the Crown Estate Office for granting permission for Tony to take photographs in restricted areas of the Great Park, along with permission to reproduce them in this format. My apologies are tendered if any unmarked photos used are known to have a particular provenance unknown to myself.

Paul Lacey   Wokingham   March 2015

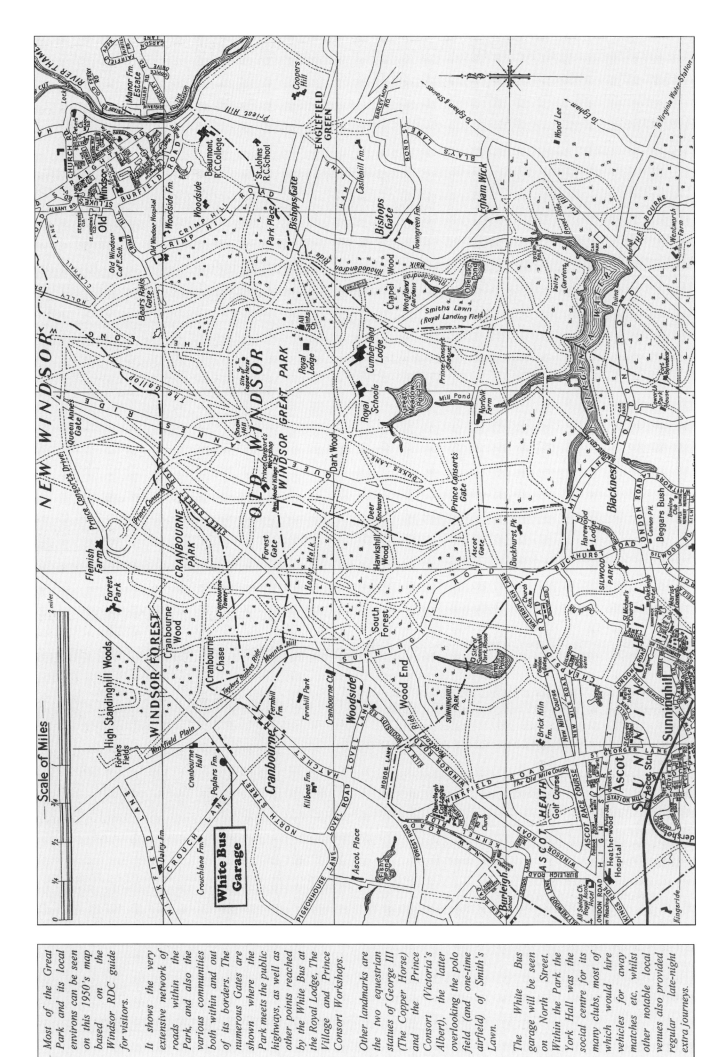

Most of the Great Park and its local environs can be seen on this 1950's map based on the Windsor RDC guide for visitors.

It shows the very extensive network of roads within the Park, and also the various communities both within and out of its borders. The numerous Gates are shown where the Park meets the public highways, as well as other points reached by the White Bus at the Royal Lodge, The Village and Prince Consort Workshops.

Other landmarks are the two equestrian statues of George III (The Copper Horse) and the Prince Consort (Victoria's Albert), the latter overlooking the polo field (and one-time airfield) of Smith's Lawn.

The White Bus garage will be seen on North Street. Within the Park the York Hall was the social centre for its many clubs, most of which would hire vehicles for away matches etc, whilst other notable local venues also provided regular late-night extra journeys.

4

Introduction to White Bus, The Great Park and Environs

The historic town of Windsor requires no introduction almost anywhere world wide, with its magnificent castle perched above the River Thames.

To the south of the castle grounds grew up a Royal hunting preserve from Norman times, now expanded to the Great Park of more recent times, which has a circumference of some 14 miles, though in earlier days the Royal Forest extended west beyond Wokingham and south to the heath-lands of Surrey.

Much of this story centres on that unique area, which has a timeless feel and has provided both work and leisure for many over the centuries, the influence of which also extended to those settlements which grew up outside the boundary of the Park itself.

Indeed the dominance of the Great Park is echoed in the names of pubs around the area, many of which will feature in the narrative as stopping places for the buses. The Royal association is reflected by the Prince Albert (Clewer Green), Prince of Wales (Winkfield), Duke of Edinburgh (Woodside), Queens Arms (Lovel Hill), Cranbourne Tower (North Ascot) and Fleur-de-Lis (Lovel Hill), whilst hunting is commemorated by the White Hart (Winkfield), Royal Hunt and Queen's Stag Hounds (both North Ascot). The workers of the forest are remembered by the Royal Foresters (west of Ascot), the Crispin (Woodside) and the Old Hatchet (Lovel Hill), whilst the nearby Ascot Race Course, with its Royal origins back to Queen Anne in the 18th century, gave the Gold Cup (North Ascot), Horse & Groom and the Nag's Head both in Ascot High Street.

One pub, opposite the centre for the *White Bus* story for many years, was the Hernes Oak in North Street, Winkfield. The legend of Herne the Hunter goes back to the 1450's, when he was gored by a stag he had injured on a hunt in the forest. His wounds duly led to madness, after which he rode his black horse with the stag's antlers fixed to his head until hanging himself from an oak tree in the Great Park. Since then his ghostly figure, along with a pack of large black hounds has been seen disappearing into the mist, often as a prelude to a dark event of history.

The maintenance of such a large hunting area, and in turn the Great Park now enjoyed by the public, required a veritable army of workers, so groups of cottages developed within and without its designated area. In particular, there are a number of settlements which owe their origins to being just outside the control of the Royal administration, whilst villages such as Winkfield and Sunninghill occupy spread out areas with other hamlets included, all of which we shall hear more of in due course.

As to distances, the North Street area of Winkfield is some 4 miles from Windsor Castle or Guildhall, whilst the centre of Winkfield parish at the church of St. Mary and the ancient White Hart pub (at one time where the Forest Court handed out its harsh punishments to poachers) is a further 1.7 miles west. To Ascot from North Street by the direct route via Plaistow Green, Brookside and Ascot Heath Crossroads to Ascot (Horse & Groom) is 3.75 miles, but by way of New Road and Fernbank Road becomes more like 5 miles. Onwards from Ascot is Sunninghill Village at 1.5 miles, whilst to Sunningdale Station is 3 miles, or to Bagshot (The Square) at 5.25 miles, all of these roads featuring in this story.

Many of the other local settlements developed around a junction, all just tracks then of course, often in green clearings in the former forest, hence local places names such as Maidens Green or Woodside. Before the days of motoring for the masses such places were really quite isolated, and life revolved around the land. Workers walked, or maybe cycled, to places of work, most people grew much of their own vegetables and fruit, whilst school children attended local schools and only until age 14 at most, so transport to school was not then the feature of daily life it would later become.

It is even quite notable that there are much less in the ranks of the Country Carriers serving Windsor, due to the large area of Great Park with its private roads, than in other comparable areas, whilst the focus for most was the weekly market at Bracknell, some 3.4 miles west of Ascot High Street.

Not that the railways had completely ignored the area, but Royal resistance in the shape of Queen Victoria (who despite her general liking for train travel), had vetoed plans for a line linking Windsor with Ascot across the Great Park as desired by the *Great Western Railway*. So, Windsor was connected to London by the *London & South Western Railway* from one station built by the riverside, whilst the *GWR* came in as a branch off the main line via Slough to the more central station opposite the castle. The latter had desired to tap into the lucrative race-going transport, though the LSWR did of course have its Ascot station linked to Waterloo via Staines and also to Aldershot, Bagshot and Guildford to the south or to westwards to Bracknell, Wokingham and Reading.

The subject of building a railway continued to be raised for many years, and even included a proposal to tunnel under some of the area, and it was not dropped by the GWR until 1911, by which time an alternative solution had been found, which we shall soon see.

And so it was to this quite singular area that the subject of a bus service would duly be developed for several quite different reasons, including the origins of the *White Bus Services*, now the only surviving independent bus operator in Berkshire which can trace its origins back to before the 1930 Road Traffic Act in a direct succession of ownership, and now in its 85th year in the care of the Jeatt family.

I use the term 'care' with purpose, as during its history the *White Bus* has at times continued to operate without really being a viable concern, the loyalty to those living in those scattered areas being given priority over profitability.

White Bus has nonetheless survived through the eras of the 1930 Road Traffic Act, the setting up of *London Transport* in 1933, then successive waves of political change encompassing Nationalisation, De-regulation and the varying involvement by Local Authorities in the provision of local bus services, for either local reasons or in response to legislation.

Also significant for this story is the Education Act 1944, which set up much of what still remains as the basis for today's home-to-school transport regime, whilst the decline of other bus services in general has given greater emphasis to such operations, whether delivered by contract or dedicated registered services.

It is into this world of change, from when motor vehicles were a rare sight and a 14-seater Ford Model T covered the service, to our modern world of private car ownership and buses with low-floor layouts and buggy or wheelchair spaces, which the story of *White Bus Services* will unfold over nine decades.

My own first contact with the firm came in 1963 when as a young bus-spotter who passed the North Street yard daily on my school journey, I decided to drop off the homeward run to investigate the yard of old buses. My school pal and I were greeted in friendly fashion and shown around a number of vehicles now disused, an OB Bedford and a Commer 'Commando' I recall.

Indeed, that *White Bus* has not only survived those often difficult times must stand as a tribute to the three generations of the Jeatt family which has for the past 85 years operated vehicles bearing that name, and it is with great pleasure that I bring you this fascinating account.

However, the story of *White Bus* actually starts 8 years before the family took over in 1930, whilst developments did not occur in a vacuum, so before the story of the firm under the Jeatts can begin we must review other operations and the family background.

This account will therefore include the pioneering services of the *Great Western Railway*, the *British Automobile Traction Co. Ltd.*, and its successor the *Thames Valley Traction Co. Ltd.*, as well as another Windsor – Winkfield – Ascot operator *Bruce Argrave* and his *Vimmy Bus Service*.

The earliest known incidence of motor buses in the Great Park captured by the camera is this gathering of four double-deckers of the Great Western Railway. The occasion saw a large group of nurses attending a Royal review by King George V of the St. John's Ambulance Brigade, which was held just off the Long Walk on Saturday 22nd June 1912. The buses shown are all Milnes-Daimler 20-28 horse-power chassis of the same type used on the Windsor – Winkfield - Ascot route, with registrations LC 6700, AF 274 and AF 157 all dating from 1905-7.

The Road Motor Service of the Great Western Railway

As noted, the *Great Western Railway* had wanted to connect Ascot with Windsor by rail, but from August 1903 it had joined the slim ranks of operators of motor buses. It did so for two reasons, either in place of the expense of constructing a permanent way, or in order to provide feeder services to railheads.

The possibility of a Road Motor Service connecting Windsor with Ascot had been considered since the Autumn of 1904, with a link from Windsor to Slough already being in operation. It is worth noting that at the close of 1904 only 243 motor vehicles of all types had been licensed in the whole of Berkshire, whereas the *GWR* bus fleet now amounted to 36 vehicles in all areas, and would in due course rise to make it one of the largest single provincial operators.

And so it was that on Wednesday 5th April 1905 the bus service commenced from Windsor (GWR Station) to Ascot (Horse & Groom), which notably did not therefore connect with the L&SWR Station a short way down the hill from Ascot.

For this new operation a new Milnes-Daimler 20hp double-decker with open-top, open-staircase double-decker body with seats for 36 was sent to the Slough allocation, under which the operation was controlled. The body had been a show model built by Christopher Dodson of Willesden exhibited at the Commercial Motor Show and registered in Buckinghamshire on 24th March as BH 02 in the Heavy Motor Car Series.

The service had 6 return journeys on weekdays and Saturdays, taking 45 minutes from Windsor to Ascot over the 7.75-mile route, and the 6-day operation was covered by the same crew.

Milnes-Daimler bus 41 (BH 02) is seen outside the Horse & Groom in Ascot High Street with the original crew of Driver Brown and Conductor Nixey, whilst the route is sign-written on the upper panels.

The bus ran out of Windsor past the Cavalry Barracks/ King Edward VII Hospital – Clewer Green (Prince Albert) – Winkfield (Squirrel Hotel) – Lovel Hill (Fleur-de-Lis Hotel) – Ascot Heath (Windsor Cross Roads) – Ascot (Royal Hotel) – Ascot (Horse & Groom).

It seems that the bus was outstationed from the start of the operation, being kept either at or very close by the Fleur-de-Lis at Lovel Hill. It is understood that the bus was periodically exchanged with another from the Slough allocation via the Windsor – Slough service as required for maintenance, and certainly a photo exists of very similar bus 61 (AF 148) working the service with route boards in place.

The first change in operation we hear of is in July 1906, when the weekday frequency was increased to 7 trips, plus an additional late-night journey on Saturdays into Windsor and back to Lovel Hill, whilst it is believed that Sunday journeys had been added that Summer, though covered by a Slough-based crew and bus instead.

It is interesting to note that local roads in those days were not sealed, consisting of rolled chippings and very dusty in dry weather and muddy in wet conditions due to a lack of drainage – the camber of them also being an additional hazard for drivers of double-deck buses! The timetables therefore carried the statement for many years that the service would run 'condition of roads and circumstances permitting' until the mid-1920's!

As was usual with the rail services, each Autumn saw a reduction in frequency, with less passengers wishing to travel in the dark, and indeed those using the buses in such times did so for pleasure, visiting or shopping, the operations not being geared to the needs of work people. Boards were erected in 1906 at various points to display the timetables and rail connections.

However, it is clear from the outset that parcel traffic was as much a source of income to the railway as its human cargo, and agencies were set up at The Squirrel, the Fleur-de-Lis and the Horse & Groom for the dispatch or receipt of parcels, which of course could be forwarded by rail all over the country.

In Royal Ascot Race Week each June the normal service was suspended during the relevant times and replaced by a shuttle service of buses <u>at higher fares</u>, with extra vehicles being drafted in via Slough Works. However, on the other hand, regular travellers could by books of 24 tickets at discount prices, and it does seem that a regular clientele of passengers did exist, and although the number of journeys did vary over the years, the basic commuter timings were maintained. Even during the Great War the service did continue with slightly reduced off-peak cover.

Soldiers in uniform were also given a discount, allowing them to travel between the Barracks and the Town Centre in Windsor at half fare of 1 penny. The other fares at the Summer of 1914 were from Windsor to Clewer Green 3 pence, to Winkfield at 5 pence, to Lovel Hill at 6 pence, to Brookside at 7 pence, to Ascot Heath Cross Roads at 8 pence, Royal Ascot Hotel or the Horse & Groom at 9 pence.

Throughout the period 1905 to 1919 the Road Motors were free from any competition, but after that the developments of the *British Automobile Traction Co. Ltd.* did have a bearing, though remarkably the service would be maintained through to 1931, when other events took place, all of which will be reviewed next.

At January 1921 the service levels were still 5 or 6 per day and still operated by the Lovel Hill outstation. At one point, the dates for which are lacking, operation of the morning and afternoon 'businessman's special' was covered by a saloon bus kept in the yard of the Horse & Groom at Ascot, and it is believed to have used the same connecting times as formerly served by the first journey from Lovel Hill to give a 9.30am train into Paddington and a 4.30pm return, but those journeys were back to starting at Lovel Hill from the timetable of May 1923. From 1922 at least the Lovel Hill-based driver was Jack Moran, who with his fine waxed moustache lodged with Mrs. England at nearby Waterloo Cottage in Hatchet Lane.

Whereas the Road Motors had featured many double-deckers in the early years, the fleet gradually became equipped with single-deck vehicles, the cambered roads and roadside trees having caused problems on rural operations. As far as Lovel Hill is concerned, a Maudslay saloon certainly took over by the war years, whilst after that high numbers of Thornycroft A-types, along with larger Guy and Maudslay buses were the standard intake.

Whereas the *Great Western Railway* had in its early days suffered from little competition, after the Great War other operators cropped up in most places, so many services were enhanced or abandoned. In the case of the Windsor – Winkfield – Ascot route the mid-1920's saw a much improved level of service over more hours of the days, itself a reflection of the better road surfaces now in place.

From February 1925 the outstation was not in use, with buses working from the Windsor end instead, and an increase to 10 return journeys, the earliest of which left Windsor at 8am for Ascot. Certainly this was in response to the improved service put on by the *Thames Valley Traction Co. Ltd.*, which in May 1924 had opened a new garage at Ascot, joining its Reading and Windsor services to form a through route and re-equipped with larger double-deckers over that road to become Route 2, a number it would retain for years.

Exactly what this GWR Thornycroft was doing at Forest Gate in Windsor Great Park is not known, but it is typical of the buses in use from the mid-1920's. Bus 909 (XY 2101) was new in 1925 with a Vickers 19-seater bus body.

In September 1925 the route returned to operation by Lovel Hill and reverted to a lesser frequency, handled in that fashion until at least July 1928, with any Sunday journeys covered by a bus from Windsor.

From 8th July 1929 an augmented service was put on, once again worked from the Windsor end only, with 3 journeys now serving the new through route via New Road and Fernbank Road, which was also being served by *Thames Valley, White Bus* and *Vimmy*. It should also be appreciated that *Thames Valley* had put its new Leyland Titan TD1 double-deckers on that road from May 1928, which greatly improved both the passenger comfort and schedules.

Under the new timetable of 1929 the 10.20pm theatre bus departure from Windsor ran through to Ascot, where it formed the 10.55pm daily departure to reach Windsor at 11.25pm, making it the latest such journey in the history of that road.

With legislation pending on road transport licensing, 1930 saw the *Great Western Railway,* in common with other bus-operating railway concerns, opting to transfer services to territorial bus companies and to take a shareholding in those concerns instead.

In the case of the Windsor – Winkfield – Ascot route, which of course had been running alongside *Thames Valley's* Windsor – Winkfield – Ascot – Reading service for many years, the transfer in 31st May 1931 merely saw the former *GWR* workings discontinued in favour of the established frequent headway. However, all buses were now operated from the Ascot Garage, so the very late Windsor bus no longer returned to that point.

The brown-and-cream liveried railway buses had run for a remarkable 26 years, despite other competitors appearing some 14 years after the first had bus set out on solid tyres over the unmade roads.

British Automobile Traction Co. Ltd., 1915 to 1920, and Thames Valley Traction Co. Ltd., 1920 to 1930

The Reading Branch of the *British Automobile Traction Co. Ltd.*, formed the nucleus of the *Thames Valley Traction Co. Ltd.*, when it became a separate entity on 20th July 1920. Although *British* expanded locally from Reading and Maidenhead from its beginnings in July 1915, wartime shortages of vehicles and particularly drivers had thwarted further developments until both were more readily available.

The road from Windsor to Ascot was first covered from July 1919, when the existing Maidenhead – Windsor service was extended onto Winkfield – Lovel Hill – Ascot Heath (Crossroads) – Ascot (Horse & Groom) – Sunningdale (Station), in fact over mostly the same route served by the Road Motors of the *Great Western Railway* since 1905. However, competition with the latter was not the intention, but to form a connection at Ascot with another service already operating from Reading – Wokingham – Bracknell – Ascot – Sunningdale, which had started in December 1915. It even featured some projections onto Virginia Water from March to October 1916, but that was ended by the Local Government (Emergency Powers) Act 1916 on the issue of damage to roads.

A bus was outstationed in a rented barn at Englemere Farm off Blythewood Lane from November 1920, with a second bus added from May 1922, and the allocation rising to 4 in due course.

BAT Thornycroft J-type Car 336 (DP 2597) in the Saxon green livery and with boards for the route from Maidenhead to Sunningdale as newly turned out.

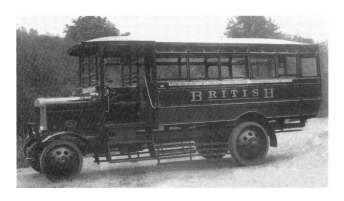

Development of *Thames Valley* routes in the 1920's is far from a straightforward matter, as the Company was never funded to any great extent, whilst other priorities and various competition saw it make changes to services quite frequently. For those reasons it is not intended to consider every amended route in the context of this review, as fuller details can be found in the dedicated volumes, but rather more to show the extent of local activity by the Company.

The next significant change came in May 1922 when the Maidenhead – Windsor section of the route was joined with the Windsor – Staines service, whilst the Windsor – Sunningdale was united as a through route with the Reading – Ascot service, still with some runs to Sunningdale Station as Route 2.

At the same time a new Route 3 was started to reach Woking, which ran from Ascot (Horse & Groom) via Sunningdale (Station) and Chobham. That lasted for exactly one year but was then discontinued.

The next stage in the development of Ascot area operations was the opening of a new garage in Course Road, just off the High Street, which opened in May 1924 and saw the 'mainline' Reading – Windsor route re-equipped with 54-seater double-deckers on rebuilt Thornycroft chassis in a new red and white livery.

Photos of the Ascot-based buses at Windsor are very scarce, but this one shows Driver Hester with Car 88 (MO 4306) standing in the High Street circa 1925.

Relations with the neighbouring *Aldershot & District* were notably cool in those days, made even more remarkable by the fact that both companies shared the same Chairman and many Directors, though there was of course an Area Agreement from early on. It was therefore of some concern that *A&D* proposed in the Spring of 1928 to run a Route 50 (Aldershot – Camberley – Bagshot – Ascot – Windsor) and Route 51 (Guildford – Chobham – Sunningdale – Ascot), to include a Dormy Shed at Windsor! It seems that the respective Directors duly agreed to avoid such moves.

Whether inspired by the above proposals or not, a new service commenced in April 1928 from Windsor to Bagshot by way of Winkfield – Ascot Heath (Crossroads) – Sunninghill – Swinley Turn – Windlesham as Route 2a, which like the previous Woking operation was covered by an Ascot-based single-decker. In that form it continued until May 1933, though the frequency and days of operation varied often.

Indeed, the section of route between Windsor and Sunninghill would continue as a new Route 2c, which will be reviewed alongside contemporary *White Bus* events in due course. There was even a Bank Holidays and Summer Sunday shuttle Route 2b from Ascot to Virginia Water (Wheatsheaf) from May during 1929.

The other relevant operations centred on the desire of *Thames Valley* to link Windsor with Bracknell by way of the 'back-roads' past Winkfield (Church) and then either through Warfield or Chavey Down, much of which also had to do with competition with *Vimmy Bus Service,* along with a growing local population.

Again, the story of developments on the back-roads corridor is quite complex, and it should also be appreciated that the Parishes of Winkfield, Warfield and Easthampstead shared a Union Workhouse built at Reeds Hill in Easthampstead, so that was one objective of such a connection. At the same time *TV* was trying to resolve how best to serve other sparsely populated areas, preferably without increasing buses and crews overall.

For the above reason, the Company is first found on the Windsor – Bracknell road through the villages in October 1929, with a rather convoluted route from Reading – Shinfield – Arborfield Cross – Barkham – Wokingham – Binfield – Bracknell – Warfield Street – Winkfield Row – Brock Hill – Maidens Green – Winkfield (Church) – Winkfield (Squirrel) – Clewer Green – Windsor (Castle)! This remarkable route was in fact covered by a Reading-based saloon as far as Bracknell, where passengers transferred to the Ascot-based saloon for the onward journey. As a 'trial route' it was not even allocated a route number.

From May 1930 the section west of Bracknell ceased as a service, with other arrangements covering the Reading – Arborfield – Wokingham and onwards to Binfield. However, the Bracknell – Windsor section remained un-numbered until May 1931, when it became the 2b and was at last extended onwards the short distance from Bracknell to Easthampstead and the Union, still covered by an Ascot-based saloon.

As can be seen from the above, *Thames Valley* made a number of changes to services in the area, whether to make connections with its territorial neighbours, fight competition or make better use of its resources.

With respect to the effects of such moves on *White Bus*, there are several particularly significant points to consider. Firstly, when their little Ford buses came along in 1922 they had an advantage in that being on pneumatic tyres they could outpace the lumbering solid-tyred Thornycrofts, though only with 14-seats.

In the earlier days the services of both probably met local need, but of course the larger concern could develop greater frequencies than the smaller operator, so some squeeze seems evident by the mid-1920's, particularly once the new Ascot Garage had opened. After that the next significant factor was the introduction in 1928 by the *'Valley* of the revolutionary Leyland Titan TD1 double-deckers, with 6-cylinder engines, covered tops and better interiors, which would have made some of the older *White Bus* and *Vimmy* stock look poor by comparison.

There had also been less direct competition initially, as the *White Bus* drew its passengers from the New Road area, which until the joining of that road with Fernbank Road, when from May 1929 the *Thames Valley* mainline route went that way. *Vimmy* also sent some journeys that way, as did the *Great Western Railway,* all of which increased the competition between the parties and, ultimately, would see *Thames Valley* reaching for the chequebook.

Another factor which also came into play was the attitude of Authorities at the three levels of Local Government as was then the case. Parish Councils often called for better services for their parishioners, though never with financial support. Rural or Urban Councils also supported the value of bus services in the trade they brought, some also being Licensing Authorities with powers under the Town Police Clauses over Hackney Carriages (which then included buses), so they could make or break such plans. But neither of those levels had direct responsibility for the upkeep of roads (and bridges which could be damaged by heavy traffic), so the County Council also had its reasons to oppose bus services or impose restrictions!

All that changed with the Road Traffic Act 1930, which came into effect from 1st January 1931. That swept away the local licensing powers and other matters concerning the routing of services, licensing of drivers and conductors, as well as introducing new legislation on vehicle design, periodic inspections and in general record-keeping. All of these areas were much more onerous for the small operator, so the Act has sometimes (unfairly perhaps) been seen to favour the larger concerns.

Therefore, after 1931 the nature of competition would change, whilst the increased paperwork, the need to replace vehicles or legal compliance would see many smaller operators sell out, as at least the new Road Service Licenses could be deemed to have a value which aided such negotiations. The territorial concerns were often aided by railway money that gave them the backup required for the purchasing spree which inevitably followed the 1930 Act.

As we shall see as the story unfolds, all of the above factors would have their bearing on the events from 1930 onwards, actually making the survival of the *White Bus Services* even more remarkable!

George Francis Ackroyd
White Bus Service of Ascot

George Francis Ackroyd was born on 8th September 1896 at Birkenshaw in the West Riding of Yorkshire, the son of a successful worsted manufacturer. By 1911 both he and his older brother Thomas were attending the famous Rugby School, at which point his parents were living comfortably at Trinity Terrace in nearby Gomersal in the 'Woollen District'.

At Rugby he attended the Officer Training Corps, and after leaving school he started with the Singer Motor Company in Coventry as a trainee, but the Great War started a month before his 18th birthday. After training at Grove Park in South London in February 1915, during April he shipped from Avonmouth Docks to France as 2nd Lieutenant in the Army Service Corps (Mechanical Transport) with the 55th then 181st MT Cos. as part of 33 Support Corps. Quite early in the war he applied for a transfer to the Royal Flying Corps, but was not successful at that point in time.

After recovering from injuries in 1916 he joined the 606th MT Co. then in 1917 was sent to Salonika in Greece with the 683rd MT Co. and Heavy Repair Workshop. Whilst there a further application to transfer for what was now the Royal Air Force was granted, after which he was sent to the RAF Expeditionary Force in Egypt, as a Temporary Lieutenant, by which time his mother was given as next-of-kin and as residing in the prestigious Cheyne Walk in Chelsea, whilst brother Thomas was also serving as a 2nd Lieutenant in the Army.

However, such service was brought to an abrupt end with the amputation of his right leg following an accident with multiple fractures, which saw him shipped home to the Central RAF Hospital in Finchley, North London. He then moved to Grenville Place in Kensington, West London, being fitted in due course with an artificial leg at Roehampton Hospital, finally resigning his commission in October 1919.

Quite why he came to North Ascot in Berkshire is not known, but before we go on with developments, it should be noted that he was known locally as 'Captain' Ackroyd although not actually of that rank. The other fact to consider is that when the military contacted his old headmaster at Rugby, the latter noted he had only ever known him as 'Francis', from which we must deduce was his preferred first name.

Frederick John Brown

Also before we get down to the development of the bus service, we will take a look at the background for Fred Brown, whose increasing role in the enterprise will soon become fully apparent.

Unlike Ackroyd, Fred Brown was of fairly local provenance, being born at Maidenhead, Berkshire in 1883, the son of a tailor's assistant. He evidently did not wish to follow that trade, whilst the family moved by 1891 to Kingston in Surrey. Instead, by 1901 we find Fred serving in the Royal Navy aboard HMS Albatross, a powerful torpedo boat destroyer launched in 1900 and built by Thornycroft at Chiswick with a length of 227ft and capable of 31.5 knots. This triple-funnelled low-built craft was a one-off and spent much time in the Mediterranean crewed by 70 men.

After his naval service he married in 1907, by which time he was a railway signalman living at Teddington, on the western outskirts of London, but by 1911 he, wife and son were at Gillingham in Kent, the 1908 birth of their son indicating a move to the Medway area by then. Once again, there is no information on why Fred Brown should come to the Ascot area.

The bus service commences

As far as can be ascertained Francis Ackroyd started the service himself, then presumably Fred Brown joined him with the addition of the second vehicle, and certainly early on they both drove on the service, despite the 'Captain' having a wooden leg.

Also, as noted, both men evidently had a practical experience of mechanics, an essential requirement with such a service to maintain. However, Ackroyd's somewhat better-off background would have a bearing in due course, as once his inheritance was settled, he put the daily operation of the bus service in Brown's hands, though still retaining the ownership of it, and we shall see these changes as the story unfolds.

Francis had come to live in New Road by July 1922, the eastern side of which had been developed with mainly pairs of semi-detached cottages, each named with a stone plaque set into the front wall, often with a date as well. That side of the road was in Sunninghill Parish, defined by the stream which, further along the road towards Brookside, still runs openly alongside the roadway. He lived at 'Klondyke Villas', almost at the southern end of the road, which at that time was not yet joined to Fern Bank Road, as it was known.

The latter road had developed from the other end, where it met London Road between Bracknell and Ascot, mostly with bungalows, often self-built, so these two populations were only separated by a short stretch of an unmade ride at right angles to both and passing the Royal Hunt pub.

So, at the time that Ackroyd moved there, those wanting to use the buses from either road had to walk out to the London Road or from the New Road end to Brookside, where both the established *Thames Valley*

Rather fortunately the early days of White Bus operation is quite well documented through advertising in the Windsor Eton & Slough Express, this page showing all known references, and these are reproduced close to original size.

WINDSOR & ASCOT
WHITE BUS SERVICE
(Proprietor - G F. Ackroyd).

Commencing Monday, 28th August

Bus will leave Brookside, Kennel Ride, for Ascot Railway Station at 8.55 to catch 9.20 to Town.

It will leave Horse & Groom (Ascot) at 9.45 to catch the 10.33 train from Windsor, and will then run as follows:

Leave Ascot —9.45 12.45 2.30 5.50
Leave Windsor—11.0 1.40 3.15 6.45

First announcement August 1922

ASCOT & WINDSOR
White Bus Service

IN spite of the extra service put on the Ascot to Windsor route by a big Company,

Mr. Ackroyd's little White Bus

will continue to run to its usual times which are as follows;

Leave ASCOT 9.45 *1.45 3.10 5.55
Leave WINDSOR . 11.30 2.30 4.0 6.35
* From New Road.

WAIT for THE LITTLE WHITE BUS

Above – 27th October 1922
Below – 24th November 1922

The Little White 'Bus

Prevents that Bumpy Feeling.

TIME TABLE.
(AS USUAL).

Leaves Ascot - * 9.45 *1.45 3.10 5.55
Leaves Windsor - 11.30 2.30 4.0 6.35
* New Road.
WEEK DAYS ONLY.

WINDSOR AND ASCOT
WHITE BUS SERVICE

Garage - THE ROYAL HOTEL, ASCOT.
Phone 364.

TIME TABLE.

ASCOT - 9.45 1.40 3.5
WINDSOR - 11.30 2.30 40
WEEK DAYS ONLY.

If Green Bus comes ——
Can White be far behind?

ROUTE AT ASCOT.—Bus starts from Royal Hotel, thence via Horse and Groom, Minnie Clare's Corner to Windsor, along usual Bus Route.

IMPORTANT.
Owing to lack of public support, the 5.45 Ascot to Windsor (Theatre Bus) will be discontinued on and after Monday, 14th May.

Change of route 11th May 1923

The LITTLE WHITE 'BUSES
May be engaged for :
SEASIDE TRIPS
(Approximate Fare, 10/6 per head)
PRIVATE PARTIES
(Approximate Fare, ordinary bus charges)
FAST LIGHT HAULAGE
(Approximate Fare on application)

TIME TABLE

SUNNINGHILL Cannon Corner		ASCOT Station		WINDSOR Castle
9.30	..	9.40	..	10.15
1.30	..	1.40	..	2.15
—	..	3.10	..	3.40
WINDSOR		ASCOT		SUNNINGHILL
11.0	..	11.30	..	11.40
2.30	..	3.0	..	—
4.0	..	4.30	..	4.40

Above – 29th June 1923
Below - No further adverts were placed until 17th February 1928

WHITE BUS SERVICE
WEEK-DAYS AND SUNDAYS.
SUNNINGHILL, ASCOT & WINDSOR
LOCAL SERVICE.
ESTABLISHED 1922.

An intermediate Bus now serves Royal Hotel, Fernbank Road, Ascot School, New Road, and Hatchet Lane, Cranbourne.

All White Buses connect Windsor G.W. and Ascot Stations.

Points served include Sunninghill Village, Marie Louise Club, Swinley Golf Club and Ascot Station.

For Time Tables or other information please apply to driver, or F. J. Brown, " Dartleigh," Ascot.

Phone Ascot 579.

and *Great Western* could be met for Windsor or Ascot or beyond.

It is not known precisely why Ackroyd decided to start the service, but we must assume he felt there was a demand for a faster and cheaper alternative, plus of course he had the necessary skills as a driver and mechanic.

On 27th July 1922 he licensed a 1-ton Ford Model T as MO 330, and it was actually a left-hand drive type, so perhaps a leftover from the war effort or an import? Upon that he had built a rather basic body with longitudinal seating for 14, with an entrance in the rear.

The rear view of MO 330 contrasts with today's low-floor access buses with five steps to reach the interior!

As can be seen from the first of the press cuttings, the service started on Monday 28th August 1922 with 4 return journeys timed to match certain train connections, which it is believed Ackroyd had gleaned by talking to his neighbours.

Although no fares are given for the original route, it is known that *Thames Valley* evidently ran extra buses to tempt passengers away, and after December 1922 used its own little Ford 'chasers' against the *White Bus*. It will of course be noted that indeed the bus service was known as the White Bus from the start, and not later as has been suggested in the past by others. The next notice takes account of lunch time for the driver, the service restarting from his home at 1.45pm!

Again, one assumes due to having his ear to the ground for local needs, from 10th November 1922 an extra bus was added for the evening for those wishing to attend the theatres on Wednesdays and Saturdays, departing from the Horse & Groom in Ascot High Street at 7.10pm and returned after performances had ended, but there is no record of when it ceased as it does not feature on the advert of only a fortnight later. Private theatre and dance trips could also be arranged when the bus was not covering the service, which ran on weekdays and Saturdays only at that time.

Despite the attentions of the *'Valley,* the new service must have proven worthwhile, indeed sufficiently so to warrant adding a second bus. That was a further 1-ton Ford T-type, though this time with right-hand drive, both of them of course being on pneumatic tyres, whereas the larger buses of the other operators were still on solids. It was registered MO 1512 on 18th May 1923, and like the earlier example was painted white with black wings.

The second Ford is seen by the junction of Ascot High Street and Station Road. It had 14 seats front-facing in a more spacious body, which although containing curved panels was still a fairly basic structure. The bodybuilder of both Fords remains unknown.

One of the most popular local events at that time was the regular dance held at the Royal Ascot Hotel, which stood to the east of the Kings Ride and London Road crossroads and opposite the race course and the more recent Heatherwood Hospital, that is until it burnt down in the 1960's. The monthly dances took place from 8pm to 1am at 5 shillings entry, but had well-known bands such as Harry Carver and his Syncopated Band in November 1922. This started the use of the *White Buses* for such late-night journeys, a feature we shall encounter again from various venues.

In a further effort to avoid clashing with the other services, the route was altered from the week of 11th

May 1923, as described in the advert on page 12, but for those not familiar with the locality, the other point was that by then the bus (and of course the second one licensed the following week) were garaged at the Royal Ascot Hotel. The service now started from that point, then proceeded down the High Street to the junction with Winkfield Road (where stood the ladies' outfitter and milliner Minnie Claire's shop) and turned left to duly rejoin the road to Windsor at the Ascot Heath Crossroads.

The reference to <u>green</u> buses does really mean the *Thames Valley,* as the red-and-white scheme did not come in until the Summer of 1924. Also, it will be noted that a further attempt to run a later theatre service had not met with the necessary support.

Fred Brown was initially living at 'Klondyke Cottages', close by to Ackroyd in New Road but a separate pair of houses, and it is understood that with the addition of the second bus he became involved with the business, presumably as the additional driver. Having another vehicle also allowed promotion of private hire, light haulage and even seaside trips!

In respect of the bus service the advert of 29th June 1923 shows that the route had been extended to serve Sunninghill (Cannon Corner) twice daily, which was probably then by way of a fairly short extension east from Ascot High Street along the A329, where the bus could turn on the triangular junction by the Cannon pub, though later we shall see that it penetrated deeper by another routing. There were still journeys taking in Ascot Station, half-mile from the High Street.

Francis Ackroyd also ran a motorcycle of unknown make, registered to him at 'Klondyke Villas' in August 1923 as MO 2109, but by June 1924 he put a car (also unknown make) on the road as MO 3381, by which time he was residing at 'Sandridge', the next house westwards from the Royal Ascot Hotel.

The service changes evidently paid off, as more seating capacity was needed in due course, resulting in the purchase of a 20-seater saloon bus body on an American-built Republic 10F-type chassis. That firm was based in Alma, Michigan and was at the time the largest truck maker in the US, whilst this model sold in Britain featured a low-slung chassis, a powerful 22.5hp Lycoming engine and good brakes! Although never sold in numbers, it is interesting to note that several similar buses were sold to operators in the nearby Woking area, suggesting some dealer activity locally, perhaps also with tempting hire-purchase?

The new bus was registered MO 5816 on 2nd July 1925, and the fact that it was painted red, white and blue also suggests it was offered as showroom stock. By then Ackroyd had relocated to 'Briar Cottage' on the London Road between the Fernbank Road turn and the Royal Ascot Hotel at the top of the hill a short walk away.

Where the entrance on the body on the Republic was, or indeed its provenance, but apparently all the *White Buses* operated as one-manners, so a front entrance would seem most likely for that capacity.

Queen Mary waves to children and nurses on her way to Ascot Races, with King George V having passed the red, white and blue-liveried Republic bus parked across the road, the rear centre exit indicating the likelihood of a front entrance. A later photo shows that this body ended up as a garden shed for the Jeatt's at North Street in Winkfield.

By 1926 Francis Ackroyd had been joined by his widowed mother Florence, and later that year she died with the result that, once her affairs were settled, it is evident that the dynamics of the business changed. Francis in fact moved into the Royal Ascot Hotel by the Autumn of 1927, but by the Spring of 1928 he had left the area to enjoy his inheritance, leaving Fred Brown to manage the bus service. Certainly, as each vehicle came up for re-taxing, it was the latter's name now given, though there is no common date to reflect any formal transfer of business, as has also been suggested previously.

Indeed, by February 1928 it is only Brown's name which appears on the publicity for the bus service, but it should be noted that it is still as the *White Bus*. As can be seen on page 12, the route had now been extended to better serve the Sunninghill area by going on past Ascot Station and the Marie Louise Club to reach Sunninghill Village, rather than Cannon Corner.

It has been suggested by another researcher that the alteration of route led to another change of garaging to The Common Garage at Sunninghill, which is logical but not documented.

By February 1928 Fred was living at 'Dartleigh' on the same stretch of London Road where Ackroyd had been in 1926. In respect of other drivers involved with

the service, Hilda Cox recalled that George Garrett (brother to Harry, of whom we shall hear more of later) and Fred Brown were the drivers but presumably after the departure of Ackroyd, from January 1927 Cecil, the son of Will Jeatt took over driving duties from Brown.

Then in July 1928 a further bus is registered in the name of Fred Brown, that being a Dennis G-type 14-seater saloon bus. Registered RX 2847 on 21st July it carried a livery of white and black, and that same day the Ford bus MO 1512 was de-licensed, although the older Ford remained in use until September 1929 after transfer to Brown's name. There is an element of mystery regarding the Dennis, which although bought for the service, disappears after a relatively short time, and it would seem that financial problems probably led to it being repossessed by the makers.

The above, along with the laying off of Cecil Jeatt in due course both point to a decline in the fortunes of the service, despite the fact that some journeys had taken advantage of the linking of New Road and Fernbank Road from early 1928, though the latter increased the competitive element with *Thames Valley, Great Western* and *Vimmy Bus Service*.

The other factor was undoubtably the legislation proposed under the Road Traffic Act which, towards the end of 1930 would convince Ackroyd and Brown it was time to sell up. Full details of sale of the bus service to Wm. Rule Jeatt will be found on page 22.

As regards Francis Ackroyd, he duly relocated to the New Forest at Balmer Lawn Road in Brockenhurst, but sadly did not have long to enjoy his days living on the proceeds of the family inheritance as on 23rd October 1938 he died at Lymington Cottage Hospital of acute appendicitis, then aged just 42.

It would seem that both of the White Bus Ford buses came to rest in the yard at Will Jeatt's garage in North Street at Winkfield. Here we see the front end of MO 330, with high-built body and over-cab luggage compartment. It is not clear if the chain etc. hanging from the Ford were once connected to it or not!

Bruce Douglas Argrave
The Vimmy Bus Service

Bruce Argrave was born in the Canterbury area of East Kent in 1870, and by 1881 we find him living with his parents at Hock House in Dunkirk, some 5 miles west of the city. He duly married Elizabeth Mitchell at East Brixton in South London in 1890, and the following April they were at 8 St. Paul's Terrace in Edmonton, North London, where he as a boot-maker.

Unfortunately, Elizabeth died young, so Bruce re-married Alice Land from North Lopham in Norfolk at Lambeth, South London in 1900, after which they had Benjamin David (1900), Walter Edward (1902) and Montague Douglas (1905), by which time the family was in Canterbury with his widowed mother at 83 Broad Street, and still there in 1911.

At some point, probably following the death of Elizabeth, he spent some time in the East Kent Rifles, though his very limited service record surviving only states he did not qualify for the Boer War medals. By 1914 the family had relocated again to Burnham, near Slough in Buckinghamshire, when he was a boot-maker. As a reservist he was duly called up but soon declared unfit for further service during 1915.

Around 1920 the family moved the short distance over the River Thames to Windsor in Berkshire, with a boot-making shop at 87 St. Leonards Road. The first we hear of his expansion of interests into a bus service comes in 22nd May 1923, when he was granted a hackney carriage license by New Windsor BC. There are no details of the vehicle he used, though a newspaper report of 1st June 1923 mentions that his 'charabanc' stopped in Windsor Forest to take a man injured in a motor accident to Windsor Hospital.

His service was known as *The Vimmy Bus Service*, and ran from Windsor (Castle) to Bracknell by way of Clewer Green (Prince Albert) – Winkfield (Squirrel) – Plaistow Green – Winkfield (Church) – Maidens Green (Crown & Anchor) – Winkfield Row (White Horse) – Chavey Down (Post Office), the last two points not then being served by any other buses. Also, of course, Bracknell then had a farmer's market each Thursday by the Hinds Head on the High Street.

The main hospital for the area was indeed in Windsor, whilst the Parishes of Winkfield, Warfield and East-hampstead shared a Union Workhouse at Reeds Hill near to Easthampstead Church. Given the latter point of interest to parishioners, it was perhaps ironic that the Easthampsead RDC objected to the appearance of the service without its prior permission. The issue was about damage to some of the roads in its care, though it did not function as a licensing authority for hackney carriages, so in September 1923 it complained to Berkshire CC, the Highway Authority in the hope that the latter would impose an order prohibiting certain roads, but that did not occur. However, it should be noted that Bruce was not the best for keeping to rules, so when he asked New Windsor for a renewal in 1924 he was reminded to keep to the conditions imposed!

Once the service was established thoughts turned to having a bungalow built at North Street in Winkfield, and between the electoral registers of 1925 and 1926, Bruce and Alice moved into 'The Bungalow', and later it was known as 'Hope Bungalow'. Sons Ben and Montague were both close-by but lodging with others, whilst the buses were now also kept locally, but no details of them or the exact location has come to light.

After the New Road – Fernbank Road roadway in North Ascot was made up for through traffic from May 1929, *Vimmy Bus Service* started on a second route from Windsor (High Street) – Ascot (Horse & Groom) via Clewer Green (Prince Albert) – Winkfield (Squirrel) – Plaistow Green – Lovel Hill (Fleur-de-Lis) – Brookside – New Road – Fernbank Road – London Road – Ascot (Royal Hotel), which now placed him in direct competition with the *GWR Road Motors, Thames Valley* and *White Bus,* all of whom had also taken advantage of the new link to serve the growing neighbourhood.

Only the second generation of Vimmy buses is known from photographs, and here we see a 1928 Reo bus (FG 4101) outside Wood's The Chemist on the High Street between the Castle and Guildhall in Windsor. The body on this is very similar to that on the White Bus Reo of the same age and registered FG 3768.

With the coming of the 1930 Road Traffic Act from January 1931, both bus routes were granted licenses, though of course the Traffic Commissioner did impose a co-ordinated timetable and fares on common points. By then several of Bruce's sons had joined him on the operation, though details are vague, but certainly Ben is known to have driven for *Thames Valley* up at High Wycombe by September 1928, whereas later on license applications used Montague's address once he had returned to living in Windsor.

Bruce Argrave also applied for a license for some excursions and tours, which were originally granted to start from Windsor (Parish Church). But from 1932 he had that amended to pick up at Lovel Hill (Fleur-de-Lis), Brookside (Post Office) and Winkfield Row (Braziers Lane). He also sought a general provision for late evening journeys (after the cessation of daily services) for those attending whist drives at the Cordes Hall in Sunninghill, but *White Bus* objected.

In April 1932 he also applied for an express carriage service between Windsor (Castle) and Ascot (Race Course) during Royal Ascot Race Week, which was granted with the standard conditions on fares and the route taken, including acceptance of the local Police ruling according to traffic conditions.

During June 1932 Argrave tried once again for a late license from the Cordes Hall to Winkfield, Windsor, Eton and Slough, but that was rejected because of other local providers, including *White Bus Services.*

In the meantime in May 1932, and due to the effects of the co-ordination he had been placed under, he asked to amend his Windsor – Ascot route to run on from Plaistow Green to Winkfield (Church) – Maidens Green (Crown & Anchor) – Brock Hill Farm – Hollies Corner – Winkfield Row (Mushroom Castle) – Chavey Down (Post Office) – Priory Road - London Road (Royal Foresters) – Ascot (Royal Hotel), but was rejected by the Traffic Commissioner, which probably left him struggling financially.

No mention of a renewal for the Windsor – Bracknell service can be found for 1932, and the above is believed to be his attempt to combine the existing two routes without the restrictions imposed upon them.

However, he never did help himself much in his dealings with officialdom, which resulted in many observations from *Thames Valley* and *White Bus* in respect of his irregular operations!

The other known photo show this 1928 Star 'Flyer' (RO 9027), which came to Vimmy BS in 1935, passing briefly to Thames Valley in due course as Car 313.

Perhaps as a foretaste of worsening finances, the North Street bungalow was sold, with a move nearby initially to 3 Jassimaine Cottages in 1933, where Montague and his wife had set up home, then again in 1934 to 27 Alexandra Road in Windsor, again with Montague and Violet. It should be noted that Windsor starting points are as quoted in individual applications, though actually within yards of each other, whilst several places on routes were incorrectly typeset in the Notices & Proceedings of the Traffic Commissioners and have been corrected.

Clearly the Argraves were looking for a way to improve takings, which led to a new application for a Windsor – Bracknell route in September 1933. Unlike the previous version this would leave Bracknell along the Maidenhead Road to the Plough & Harrow at Warfield Street, before turning east to Hollies Corner – Brock Hill – Maidens Green (Crown & Anchor) – Winkfield (Church) – Plaistow Green and onwards to Windsor. The license was granted in modified form, with some of the existing Windsor – Plaistow Green journeys re-timed to run on to Bracknell and back.

However, by then the *'Valley* could sense the weak position of *Vimmy,* so its Inspectors kept a note of bus timings, which resulted in the section from Plaistow Green to Bracknell being revoked from 16th April 1934! At the subsequent appeal hearing Bruce Argrave stated that he employed 7 people to run the services, but the decision was upheld. *Thames Valley* then stepped in with a service from Windsor through to Bracknell, though via Chavey instead of Warfield, from 2nd June 1934 as Route 2b, extending the service onto Easthampstead (Union) one year later.

The *'Valley* had been busy with acquisitions in other areas, but for early 1936 it got around to reviewing issues in the Ascot/Winkfield area. As Bruce Argrave was now 66 years old and not in the best of health, he decided to accept the offer for his operations and two Star 'Flyer' buses (RO 9027 and VG 1631), with the transfer dated 13th May 1936 – in fact the same day that the original *White Bus* route was also taken over.

The Traffic Commissioner was understandably happy to allow transfer of licenses to *Thames Valley*, whilst the journeys were merely absorbed into the existing Windsor – Ascot – Reading route, and the race-days express added to the service already inherited from the *GWR Road Motors. TV* was not interested in the late runs from dances etc., so did not object to *White Bus* when it added such facilities to serve late-night hall users shortly afterwards.

Bruce Argrave then returned to his former trade at 20 Dedworth Road, also becoming the local parcel agent for *Thames Valley*, finally passing away at the age of 86 in 1956. The Stars were numbered 312 (VG 1631) and 313 (RO 9027) but both were sold later in 1936.

The Jeatt Family - From Origins to Winkfield

The line of the Jeatt family of interest to our story can be traced back to the 1770's when Richard Jeatt married Florence Rule, the originator of a name often passed on as a Christian name down the years. Their son Richard was born at Kingswear in South Devon in 1777, and he and his wife Rebecca had son William Rule Jeatt (the first of three to be so named over the years) in 1813 at Dartmouth, just on the opposite bank of the River Dart.

William married in Dorset, having evidently entered the Customs & Excise Service as a coast-based officer, and after being in West Lulworth in 1851, ten years later we find him as Chief Officer Coast Guard on the Isle of Portland a few miles along the coast. A son of the same name was amongst his offspring, but not of direct significance to the developments we are reviewing.

However, the latter's brother Arthur James Jeatt was born at West Lulworth in 1851, initially marrying in that village, though his wife died after they had their first two children. Arthur then re-married locally in 1878, but by 1881 the family had relocated to Arreton on the Isle of Wight, then part of Hampshire. In 1883 the third William Rule Jeatt was born to Arthur and Emma, and this was indeed the one whose life we will now follow. The couple were both employed by the County as infant school teachers and lived at School House in Arreton Street. William also had brothers named Cecil and Douglas, both names that will crop up again in subsequent generations.

By 1901 the 18-year old Billy (or later Will) could be found still living at home and working as a bicycle repairer. Soon after that we have the first reference to his involvement with motor cars, this photo dated 1902 without a location known, but on the Island.

Telegrams: "Jeatt, Puckeridge."
Telephone 7 Puckeridge.
Station: Stanton, G.E.R.

Wm. RULE JEATT,

MOTOR FACTOR & SPECIALIST,

PUCKERIDGE, HERTS.

Agent for the " Maxwell " and " Mass-Paige " Cars.

Any make of Cars supplied.   Second-Hand Cars taken in part payment.
All makes of Tyres & Accessories supplied at London Prices.

Above – *A business card from the Puckeridge garage.*
Below – *That garage as seen in a postcard of the High Street in the village, noting the signs also for cycle repairs and accessories.*

He first became involved with cars on the Isle of Wight, but in 1905 he was at Puckeridge near Ware in Hertfordshire, and certainly by 1913 the garage was in partnership with a Mr. Martin. The name of Wheeler's Motor Garage also appears on some documents, which may have been a reference to a previous owner or his initial employer in that area. During 1908 he married locally to Ella Paice, and they had son Cecil Edouard in 1909, followed by daughter Vivien Mary in 1911.

Records showing the partnership with Martin exist through to at least 1917, but in February 1915 it came the turn of Will to be called up for the Army. After attestation he came home until sent for in November, being assigned to the Army Service Corps as a Motor Lorry Driver, such men needing the skills of a motor mechanic in those often difficult operating conditions. He spent all his war years in France and Flanders and mostly driving munitions and supplies to the Front.

William Rule Jeatt when in the Army Service Corps.

Such work was often very dangerous, as much of it took place on unlit roads at night and under the constant threat of bombardment by the enemy. Indeed his diary is peppered with references to 'hot time from Fritz' and the likes. He was also horrified to return to the railhead after a night's driving to see some 200 horses slaughtered by an attack by heavy shells. The exact places he served are not easy to trace, as names quoted are either colloquial versions or perhaps not too specific for reasons of security, but he certainly drove for the 189th Heavy Artillery, the 290th Battery, the 8th Siege Corps and No.5 ASC (MT) Company.

In an interesting episode involving both the garage and the war, the couple received a letter in September 1918 from a Lieutenant Eric Sindlehurst, recalling the night in July 1915 when he had called in with his motorcycle, the lights of which had failed. Will tried hard to fix them, but had to concede defeat, so the couple put the soldier up for the night, gave him breakfast and he left in the daylight for Chelmsford. He had said he would send them a regimental badge from the Royal Engineers, which he now finally did, along with a Royal Warwickshire, obviously a nod to Ella's origins at Temple Balsall in that County. He added that despite his wanderings over many fronts he had never forgotten their kindness.

Will was evidently injured right at the close of the war, as a telegram to Ella at Puckeridge noted he was now in a hospital at Manchester in November 1918, before transferring to Chester in January 1919, whilst later correspondence on his war pension claim notes he had a shrapnel wound to his face.

Whilst he had been away on military service Ella had tried to continue the garage business, but by his return financial difficulties had arisen. Then, with such sad timing, Ella contracted the deadly Spanish 'flu which has swept Europe in 1919. Although it appeared that she had overcome the effects, she subsequently died in Southwark, South London later that year.

As his own health also placed him under medical supervision through to November 1920, Will decided to sell the business, though out of the £2200 he got for the garage and the goodwill, only £450 was left after outstanding debts were settled.

Clearly, with such financial problems, his two small children and still affected by war service, he was soon to re-marry in the Summer of 1920 to Beatrice Lavinia Goose at Hackney in Middlesex. She had been born at Norwich in Norfolk in 1886, and by 1911 was a nurse for the Metropolitan Asylum's Board at the North Western Hospital in Lawn Road, Hampstead, North London, and is seems quite possible that the couple met whilst Will was recovering at wherever she was then employed.

At first the couple, with Will's children, were all at Puckeridge whilst his affairs were being sorted out, and stamps in his war pension book show they remained there until late June 1921, though some 3 weeks later he was using Winkfield Post Office for his claim. In the mists of time no one can now recall quite why they had chosen Winkfield to relocate to!

Will inevitably tried to re-establish his line of business, renting a wooden workshop opposite the Fleur-de-Lis pub at Lovel Hill from Mrs. Dudley Charles of Winkfield Lodge, and about 1 mile from North Street in Winkfield. From there he undertook cycle and motor repairs and traded under his wife's name B. L. Jeatt due to his earlier financial issues.

The Lovel Hill premises with Ted Payne at the door.

Facilities at the Lovel Hill premises were fine for bike repairs, but rather limited for working on cars, so after about a year the opportunity came up to buy or lease a former rifle range in North Street, Winkfield. It was a very large tin shed, which was soon transformed into a motor garage in keeping with post-war demands.

Daughter Vivien Jeatt stands in the entrance to the old rifle range, now known as Winkfield Garage. To her rear is a large hire car with fold-down roof (?N 8323), which carries a Hackney Carriage Plate for 6 passengers. Note the contemporary adverts, range of services and brands of petrol and cars.

Living next door to the garage premises was young Hilda Cox, who as these notes are written in 2014 attained her 100th birthday, and we have her to thank for a first-hand account of developments around those early years of the Jeatts at Winkfield. The following is taken from her notes, largely as written, but items in brackets have been added where appropriate for clarification –

'The long tin building of the former rifle range was bought by Mr. Jeatt, and with the help of his brother-in-law Sid Goose it was made into a motor garage. My father excavated the hole for the fuel tank, which was itself quite a novelty locally.

Whilst the alterations were going on Mr. Jeatt and Sid stayed at The Squirrel Hotel (very close by on North Street), but Sid went back to his parent's home in Watford when the work was done. At that point Mrs. Jeatt, who we called 'Nurse', came down with Cecil, and he and I became friends. I didn't like her bossy ways and she treated Cecil badly, hitting him with a wet towel. Once he started school in Windsor he was expected to walk there (some 4 miles, and despite there being several bus services) in all weathers, so after he got sores on his face and ears due to the cold my mother knitted him a balaclava and mittens.

To make his life even harder he had to sleep on a camp bed in his father and mother-in-law's room at The Squirrel, and not allowed in the room much of the time, so he soon took to going to the garage to see Harry Garrett (born 1906 and lived at Forest Park Stables, where his father was the coachman in 1911) and Bill Peters who were now working there as mechanic-drivers, or otherwise round to see me and my parents, partly because he was always underfed, which they put right. Cecil and I often roamed the local area together, especially Windsor Forest, where armed with sweets and comics we took what hours of

childhood pleasure we could away from parental control.

To improve the accommodation for the family a tin-shack bungalow was obtained at Maidens Green, some 2.4 miles west of North Street, though it only had a Valor oil stove for both cooking and heating (known as The Bungalow).

In the meantime Cecil's sister Viv had been living with her grandparents on the Isle of Wight, but after payments for her upkeep ceased to arrive, she had been brought up to Berkshire. She also had a hard time of it as the regime in the household was austere there, and she now found herself having to sleep on a camp bed behind a screen in the sitting room, whilst she was required to do lots of daily cleaning.

After a while Mr. Jeatt had a partner in the business, a relative named Hodder, whose father had given him the money to invest (25-year old Donald Hodder from Abbotsbury in Dorset, the title shown as Wm. Jeatt & Co., Winkfield Garage from July 1924). However, once Hodder realised how the business was running there was a big bust up (in August 1925 and the partnership was finally dissolved in early 1926).'

The Winkfield Garage as seen from The Old Forge, with BP petrol signs displayed at all angles. However, that on the roof probably covers a leak rather hoping to attract low-flying aircraft using the nearby strip.

'Even before they moved from the old bungalow there was money trouble, and one night a lot of bikes from stock were moved in the dark from the garage to a store at my parent's house. Unfortunately, Will had a habit of not finishing jobs as promised, and when sought out by customers could often be found in the nearby Hernes Oak pub – not that he drank much, but he liked the company and the gossip.

Early on they had a large old Renault car (obtained in August 1921 from a London dealer and quite likely the one seen in the photo above left), which he used to take people about in, the hood went up and there was a speaking-tube from the passenger section to the chauffeur. He often took old ladies from the Marie Louise Club to Sunninghill Church, but he wasn't known for his reliability! Cecil also had a regular job at 'Laurel Dean' twice a week to run an engine used for generating electricity to charge up batteries.

The Old Forge, which was opposite the Winkfield Garage in North Street, became vacant and was taken over by the Jeatts, which meant that a lot of nice furniture which had been stored up at Ware could now come down. Even after the move there (about December 1923 it seems), the children were not allowed in their rooms except for sleeping, so Vivie spent much of her time in the garage.'

The old Renault car was converted for use as a breakdown truck, presumably by the mid-1920's as the signage indicates this was during the partnership period with Donald Hodder. It is seen outside the Old Forge, now known as North Street Garage.

'Cecil got a job driving a grocer's van for Mr. Nutley nearby, but wasn't particularly reliable so he got the sack. The family business was going from bad to worse, so 'Nurse' rented out a couple of rooms and started to keep chickens, as well as doing some part-time nursing work. The couple did have a baby on the way, but sadly it was stillborn in May 1926.

The money troubles were coming to a head, so various goods and chattels, including nice pieces of furniture were transferred to my parent's house. 'Nurse' had also fallen out with my mother, so she took a job in a hospital in Hillingdon, Middlesex. I was at that time renting a room from the man next door, whose wife was in hospital, so after the above events Mr. Jeatt rented the house and Vivie and Cecil came there, but on camp beds once again, until the house in Crouch Lane (a short way the to the west of the North Street premises) could be arranged.

Mr. Jeatt's problems were not helped by his natural generosity, as he would treat the local children to bars of chocolate or even shoes, though he was not

that kind to his own children, particularly as marrying Beatrice had driven them apart. Later on Mr. Jeatt came as lodger to me and my husband Maurice Nugent (they married in early 1931, and Maurice duly worked on White Bus), however his chronic bronchitis was a nuisance, so he went back to Crouch Lane and bought a bed which folded into a cupboard.'

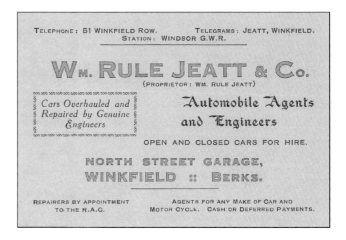

A business card from the partnership era, after the move to the former forge of Blacksmith W. Tombs, who had been there from at least 1911 to 1920.

And there we shall leave Hilda's account for now, with more information on some of the events covered. As noted, the business affairs of Will Jeatt were far from a smooth ride, but as author I must perhaps make a comment here. Having read virtually all the letters he wrote throughout the 1930's, it is readily apparent that he was an intelligent and perceptive man, so his misfortunes did not stem from any lacking in those areas. However, my Father recalls that when we came down from London to Bracknell in 1957, he found that many of the local garages owners were still letting their local customers run up accounts without effective procedures for settlement, whilst others took money for housekeeping etc. straight out of the till!

For some years a calendar was given away to regular customers, and from the 1922 version we can glean the following services were available– RAC appointed repairer, any make of car supplied, secondhand cars taken in part-exchange, open and closed cars for hire, stockists of Firestone tyres, Vacuum, Price's and Monogram lubricating oils and greases, petrol, all types of accessories for the motorist, agent for JM shock absorbers, accumulators charged, first class vulcanising, garage open day or night, inspection pit, cars overhauled and estimates free.

In order to promote the garage locally Will Jeatt took an advert panel on the staircase of one of the *Thames Valley* Brush-bodied Thornycroft double-deckers on the Windsor – Ascot – Reading route, from September 1924 at 6 shillings and 6 pence per week.

A mortgage had in fact been taken out on The Old Forge in November 1925, but the financial position was yet again reaching critical, leading to bankruptcy proceedings in October 1926. A sale of effects, based on a detailed catalogue produced by the agents, was held on Tuesday 7th December 1926 at the Star & Garter Hotel in Windsor. Up for disposal was the North Street Garage (sometimes referred to as The Forge Garage), which now comprised of a workshop 37ft 9ins by 13ft 9ins, plus adjoining sheds and old stables now used for storage, 2 petrol pumps and fuel tanks. He evidently still had ownership of the old rifle range building, which was rented out at £1 per month, and that was confirmed as 32ft wide and 40ft in depth. Also included was the goodwill of the business, but not tools and other associated equipment, which was an interesting situation.

Harry Garrett and the Napier car, showing the style of bodywork that had evolved from the private carriages of the horse era. Note the elaborate bulb-horn and quick-change wire-spoked wheels, whilst the smart leather uniform and dust-pan-and-brush all speak of a pride in the job.

After the debtors were satisfied it seems that Will just carried on by trading under Beatrice's name once again, and continuing to offer motor repairs, car and lorry hire and a breakdown service! The lorry hire was quite regular, with some local farmers and businesses preferring it to ownership, whilst the idea of moving house by one's own effort is nothing new – though it could then be a case of a 'moon-light flit' for those

with rent arrears, as my Grandad found with his hand-cart which he hired out at night to neighbours!

The garage accounts for 1929 show that *Fred Brown* was not only buying his fuel and oil there (as indeed was *Bruce Argrave* of the *Vimmy Bus Service),* but also had the remaining Ford and the Republic bus worked on by Will. The latter also delivered parcels locally for Lawrence's Stores of Bracknell, Ascot etc. that year, probably driven about by Cecil.

Also in the accounts are a lorry hire for Miss Ackroyd from Wokingham to London in October 1929, whilst the final reference to *Brown* was his own removal of house to Harrow in Middlesex in October 1930. By March he was running a Vauxhall car, but by May that year he had a Wolseley instead on the account.

In respect of cars for hire by Jeatt, the landaulette 6-seater already noted was probably the Renault which became the breakdown wagon, whilst in July 1924 a Citroen 4-seater was acquired, and around that time the Napier. A Maxwell 4-seater tourer (XN 1546) was evident in May 1928, whilst on the lorry side a Ford 1-tonner truck had been purchased in January 1923, and a late 1924 unknown model. Most of those purchased came from the same London dealer, William Bailey in Great Portland Street, along with new Maxwell and Studebaker cars ordered on behalf of clients, for which a commission provided income, as did an agency for motor insurance policies. One other amusing job was the refurbishment and re-painting in lacquer paint of a perambulator, whilst the practical skills of welding were useful for repairs.

Vivien is seen in workshop coat astride a motorcycle of late 1923 origins (XP 9083), whilst in the background is a hire lorry of unknown make (MF 8912). In due course there are occasions when she rode or drove over to Guildford to collect parts from the Dennis Bros. works in Woodbridge Road, though many other parts came via passenger train to Ascot.

Another shot of the Napier hire car (?? 7092) with wedding ribbons and being driven by Harry Garrett circa 1925.

After Cecil Jeatt had left the employ of the grocer, he actually went as driver-mechanic for *Fred Brown,* which could be said to be the start of the family's involvement with the *White Bus Service,* and certainly it must have had a bearing on events to come. Cecil joined them in January 1927, probably when the daily control passed into *Brown's* hands after *Francis Ackroyd* moved out of the area.

In fact it seems that when one of the *White Bus* Ford buses was withdrawn in July 1928 it went no further than the yard of North Street Garage, after which it presumably provided spare parts, and in due course the other bus of that make would also wind up there.

This photo of White Bus Ford T-type 14-seater bus (MO 330) was taken in North Street opposite the Hernes Oak pub, presumably when driven by Cecil.

We have already seen how the dynamics of the *White Bus* business had evolved with the changes of fortune for *Francis Ackroyd,* whilst the impending legislation of the 1930 Road Traffic Act no doubt made both he and *Fred Brown* reflect on the future of their service, which most likely had not been faring too well since

other services competed in the New Road area. *Brown* had been forced to let Cecil go in August 1929 due to such difficulties, whilst also perhaps having to default on payments for the Dennis bus. It is apparent that whilst the buses had been put under *Brown's* name (if not perhaps his actual ownership), the service had remained the personal property of *Ackroyd.*

The new legislation would come into effect on 1st January 1931, and on 16th December 1930 William Rule Jeatt signed a Purchase Agreement for the bus service and the Republic motor bus. In this document the operation is referred to as the *Republic Bus Service,* though there are no other instances of that title in adverts or local directories. As the bus of that make had been new in July 1925 any change of name should have shown up in known adverts of 1927-8. Most likely it was the fact that only the Republic now remained in use which caused that title to be used, whilst the lack of signature by *Fred Brown* also seems to confirm that he had departed. At that time *Francis Ackroyd* was residing in the Ashburton Hotel in Calstock, Cornwall, on the River Tamar inland from Plymouth. The agreed price was £100 for the bus service and £50 for the Republic bus (MO 5816), which a letter does acknowledge as 'the remaining bus in *Brown's* possession'. Payment was scheduled as £20 on 31st March 1931, followed by £30 3 months later, then £8 per month until a final sum of £4.

It is interesting that in Hilda Cox's account, in which it is stated how little she knew about *White Bus* before it came to the Jeatts, it is specifically noted that she believes *Ackroyd* had 'a debt with the garage' which actually formed part of the transfer cost.

However, the financial troubles were not yet over, as Beatrice defaulted on the mortgage payments for The Old Forge, leading to another disposal sale to be held at those premises on Tuesday 8th December 1931. Up for sale now was the Stock-in-Trade of a Garage Proprietor, 2x500-gallon steel fuel tanks, a Republic 2-ton chassis (its bus body known to have become the garden shed), several old cars, a BSA motorcycle, a 1921 27hp Buick touring car, a 1923 15hp Wolseley touring car, a 10cwt van body, 3 derelict cars, 4 old car bodies, a Wolseley 2-seater body off a 10hp car, as well as lots of Ford and Chevrolet spares.

This did indeed finally put pay to the garage side of the business, leaving Will to now concentrate solely on the bus service he had acquired, with of course the assistance of his son Cecil and daughter Vivien. The plot of land to the right of The Old Forge was now rented as a new base for the venture, a case of the phoenix arising from the ashes for the third time! That same site is still in use to this day 85 years and three generations later, developments we shall now consider. Also from now on the relevant operations of other local operators will be included in the main text.

1931 - Under New Ownership

Although the 1930 Road Traffic Act came into effect on 1st January 1931, the large volume of licensing of services to be dealt with meant that operators were to continue as before until it came to their turn for the hearing of applications. The same applied to personal licenses for conductors and drivers, along with the inspections of vehicles under the new regulations.

Indeed, Will Jeatt had written to New Windsor BC in early January to inform the licensing officer that he had purchased the service from Ackroyd, and that he was continuing to use the Republic on its then current hackney carriage license.

He also sent off his application for a Road Service License to the South East Area Traffic Commissioner in Reading during May 1931 for the continuation of his Sunninghill – Ascot – Winkfield - Windsor route, the application showing the timings to be –

		a		a		x
Sunninghill (New Inn)	8.50		1.50		5.20	7.10	9.30
Marie Louise Club	8.55		1.55		5.25	7.15	9.35
Ascot Station (S.R.)	9.3	10.32	2.3	3.32	5.33	7.23	9.43
Ascot (H. & Groom)	9.6	10.35	2.6	3.35	5.36	7.26	9.46
Royal Hotel	9.8	10.37	2.8	3.37	5.38	7.28	9.48
Royal Hunt	9.13	10.42	2.13	3.42	5.43	7.33	9.53
Brookside	9.17	10.46	2.17	3.46	5.47	7.37	9.57
Lovel Hill (Fleur-d-L)	9.19	10.48	2.19	3.48	5.49	7.39	9.59
Winkfield (Squirrel)	9.24	10.53	2.24	3.53	5.54	7.44	10.4
Windsor (Hospital)	9.33	11.2	2.33	4.2	6.3	7.53	10.13
Windsor (Castle)	9.38	11.7	2.38	4.7	6.8	7.58	10.18

			a		a		x
Windsor (Castle)	9.50	11.15	2.45	4.15	6.15	8.40	10.25
Windsor (Hospital)	9.55	11.20	2.50	4.20	6.20	8.45	10.30
Winkfield (Squirrel)	10.4	11.29	2.59	4.29	6.29	8.54	10.39
Lovel Hill (Fleur-d-L)	10.9	11.34	3.4	4.34	6.34	8.59	10.44
Brookside	10.11	11.36	3.6	4.36	6.36	9.1	10.46
Royal Hunt	10.15	11.40	3.10	4.40	6.40	9.5	10.50
Royal Hotel	10.20	11.45	3.15	4.45	6.45	9.10	10.55
Ascot (H. & Groom)	10.22	11.47	3.17	4.47	6.47	9.12	10.57
Ascot Station (S.R.)	10.25	11.50	3.20	4.50	6.50	9.15	11.0
Marie Louise Club		11.58		4.58	6.58	9.23	11.8
Sunninghill (New Inn)		12.3		5.3	7.3	9.28	11.13

Notes: a - via Cranbourne Church, x – Weds./Sats./Suns. only.

Single Fares: Windsor – Winkfield 5 pence, Windsor – Horse & Groom 1 shilling, Windsor – Sunninghill 1 shilling 1 penny.
Return Fares: Windsor – Winkfield 8 pence, Windsor – Horse & Groom 1 shilling 6 pence, Windsor – Sunninghill 1 shilling 9 pence.

It should be noted that Wednesday was the general early-closing day for shops in the district, so many used that afternoon for local visits and excursions. The above application was granted on 19th June 1931 in the form shown above, but after the hearing a letter came from *Thames Valley* noting that should in the future Mr. Jeatt's children wish to dispose of the service, it would appreciate first refusal to purchase, and indeed he responded that would be the case. Two things are notable from this, firstly that he was

certainly still head of the outfit, though it now bore the title *C.E. & V.M. Jeatt (t/a White Bus Services)*, whilst also very obvious is the civil tone in exchanges of correspondence despite the rather David-and-Goliath nature of the respective concerns.

We have already seen that the Republic had been no stranger to Will before the takeover, but unfortunately during late June 1931 it suffered a mechanical issue which could not be resolved. The situation was no doubt made worse by the fact that the type had only sold in very small numbers in the UK, whilst the 1929 Wall Street Crash had put the Republic Motor Truck Company of Alma, Michigan out of business, so it must be assumed that spare parts could not be found.

There then followed a lot of correspondence with dealers in a search for a replacement bus, and the dealer E. Wilson Blaxton of Swallow Street, Piccadilly, London W1, automobile engineers and Daimler agents put forward 3 buses for inspection at its Waterloo premises.

There was a 1927 14-seater Chevrolet (PW 9949) with front-entrance body by Waveney and a centre-rear emergency door, which was in good condition but needed a new 1st and 2nd gear, which they offered at just £30.

There are no known photos of the little Chevrolet with White Bus, but this shows one of that batch in service with United Automobile Services, who used them for 'chasing' duties for just 2 years before disposal.

The second was a 20-seater US-built Reo 'Pullman Junior' new in 1928 (FG 3768), which featured 4-wheel braking and a 6-cylinder engine, and with two entrances both on the nearside. It had not been used for a year as it stood at the Daimler works after being taken in part-exchange, a common situation in those days, where manufacturers took old stock to encourage sales, but then disposed of them via dealers away from the area of origin, hence this Fife-registered bus coming south.

The third bus on offer was another relative rarity, being a 1928 French-built Renault NY-type with 20-seater body (CN 3651), evidently hailing from the north-east and probably for similar reasons as the Reo, which went to *Comfy BS* of Horsham instead.

With the Chevrolet so cheap but repairable, Will took it and the Reo, but not the Renault, the 14-seater he worked on himself, whilst the Reo required some work to meet the new Construction & Use Regulations. As it was, he wanted the front doorway, which had been intended originally as an emergency exit to be used to facilitate one-man operation, with a door now fitted to the Scottish-style rear open platform which also featured on some single-deckers north of the border. The front doorway needed widening, whilst the rear entrance was fitted with a folding door workable from inside or out to conform. However, in order to meet the requirements, the seat to the left of the driver was removed, whilst those over the wheel arches were re-arranged, leaving the seating capacity reduced to just 16.

This view of the Reo (FG 3768) shows it decorated for the Silver Jubilee in 1935, but also clearly both the entrances. Little it known of the early history of this bus, but worth noting is that Bruce Argrave also had a Reo (FG 4104), the body of which was identical on the offside to this example, both by an unknown firm.

Work on the Reo was not completed until 4th August 1931, but Will got on and put the Chevrolet on the road for July. It was of the LM-type with American-designed 4-cylinder 2.8 litre petrol engine on a model of chassis assembled at Hendon in North London. As a postscript to its acquisition, in due course the local Inspecting Officer for the Traffic Area asked to see the Certificate of Fitness for it, so Will referred him to the dealer, who in turn asked him to contact the former owner, W.F. Alexander (*Comfy Bus Service*) of Horsham, who had to admit he had not got around to getting one before deciding to sell it!

As to the Reo, these chassis were imported from the US and had been a popular choice with independents for both bus and coach work, being fast machines but at least they had brakes to match.

Whether a standard livery was adopted from the outset for the above pair is not known, but evidently it was not long before they bore a brown-and-white scheme as can be seen from the photo above.

As to the Republic, the body was taken off and a photo shows it became the garden shed for the family, whilst the chassis is of course listed in the articles for sale on page 22. Obviously, the need to have more than one vehicle for use had been rather brought home by its sudden demise, but it also gave opportunities to develop some private hire, as well as local journeys after late-night finishes for dances etc.

Cecil had of course been working for Fred Brown on the bus as a driver, and after the 1930 Act he obtained both Driver and Conductor Licenses, as did his father, whilst Maurice Nugent (who had married Hilda Cox in 1931) also did some duties from time to time from those early days and had been born locally in 1907.

1932 – A New Link Provided

Competition on the Sunninghill – Ascot – Winkfield – Windsor route was of course now a thing of the past, but the co-ordinated timetables and common fares imposed by the new licensing regime did mean there was little leeway for expansion on that road. This set in motion a chain of events which ultimately would lead to the iconic association of *White Bus Services* and the Great Park.

The new route set out to serve the numerous small settlements which ringed the southern boundary of the Park, whilst also providing the first buses to operate within striking distance of that unique area. This latter feature would attract the attention in due course of the new Deputy Surveyor of Parks & Woods appointed by the Crown Commissioners in 1931.

The new service from January 1932 ran as Windsor (Castle) – The Fountain – Queen Anne's Gate – Forest Gate – Crispin Hotel – Woodside (Duke of Edinburgh) – Sunninghill Park – Sawyers Gate – Cheapside (The Thatch) – Blacknest – Sunninghill (The Cannon) – Sunninghill (High Street), though not as has been suggested actually entering the Park, there being virtually no habitation between Queen Anne's Gate and Forest Gate then. The timings also confirm that no such deviation was included.

The new service only ran on Wednesdays, Saturdays and Sundays, with an afternoon and evening operation aimed at the early-closing day and weekend travel for shopping, visiting or other popular leisure activities such as the cinema, theatre or boating on the Thames. Paid holidays were of course in short supply in those days, and for many Saturday (or at least the morning) formed part of the normal working week, so although people then had less leisure and sophisticated outlets for their time, they certainly knew how to exploit the opportunities open to them, whilst a walk-out with your sweetheart was often the only un-supervised time spent alone, making the Great Park a popular place for such a stroll or a picnic.

WHITE BUS SERVICES.   TIME TABLE.

From WINDSOR CASTLE to SUNNINGHILL via CHEAPSIDE.

					p.m.	p.m.	p.m.	p.m.
WINDSOR CASTLE	....	....	.... dep.	....	1.50	3.50	5.50	7.50
THE FOUNTAIN	....	....	.... ,,	....	1.55	3.55	5.55	7.55
QUEEN ANNE'S GATE	....	....	.... ,,	....	1.57	3.57	5.57	7.57
FOREST GATE	....	....	.... ,,	....	2.4	4.4	6.4	8.4
CRISPEN HOTEL	....	....	.... ,,	....	2.7	4.7	6.7	8.7
WOODSIDE (D. of E.)	....	....	.... ,,	....	2.9	4.9	6.9	8.9
SUNNINGHILL PARK	....	....	.... ,,	....	2.11	4.11	6.11	8.11
SAWYER'S GATE	....	....	.... ,,	....	2.14	4.14	6.14	8.14
CHEAPSIDE (Thatch)	....	....	.... ,,	....	2.18	4.18	6.18	8.18
BLACKNEST	....	....	.... ,,	....	2.23	4.23	6.23	8.23
SUNNINGHILL (Cannon)	....	....	.... ,,	....	2.27	4.27	6.27	8.27
SUNNINGHILL	....	....	.... arr.	....	2.30	4.30	6.30	8.30

From SUNNINGHILL to WINDSOR CASTLE via CHEAPSIDE.

					p.m.	p.m.	p.m.	p.m.
SUNNINGHILL	....	....	.... dep.	....	3.0	5.0	7.0	8.35
SUNNINGHILL (Cannon)	....	....	.... ,,	....	3.3	5.3	7.3	8.38
BLACKNEST	....	....	.... ,,	....	3.7	5.7	7.7	8.42
CHEAPSIDE (Thatch)	....	....	.... ,,	....	3.12	5.12	7.12	8.47
SAWYER'S GATE	....	....	.... ,,	....	3.16	5.16	7.16	8.51
SUNNINGHILL PARK	....	....	.... ,,	....	3.19	5.19	7.19	8.54
WOODSIDE (D. of E.)	....	....	.... ,,	....	3.21	5.21	7.21	8.56
CRISPEN HOTEL	....	....	.... ,,	....	3.24	5.24	7.24	8.59
FOREST GATE	....	....	.... ,,	....	3.26	5.26	7.26	9.1
QUEEN ANNE'S GATE	....	....	.... ,,	....	3.33	5.33	7.33	9.8
THE FOUNTAIN	....	....	.... ,,	....	3.35	5.55	7.35	9.10
WINDSOR CASTLE	....	....	.... arr.	....	3.40	5.40	7.40	9.15

WEDNESDAYS, FRIDAYS and SATURDAYS ONLY.

Above – The timetable for the new link via Cheapside introduced from January 1932.
Below – The slightly modified timetable for the original Windsor – Sunninghill link as from July 1932.
Both timetables are shown at about two-thirds original size and were black type on white paper.

WHITE BUS SERVICES.   TIME TABLE.

From SUNNINGHILL (New Inn) to WINDSOR CASTLE.

				a.m.	a.m.	p.m.	p.m.	p.m.	p.m.	p.m.	p.m.
SUNNINGHILL (New Inn)	...	... dep.		8 50	...	12 50	1 50	...	5 20	7 20	x 9 30
MARIE LOUISE CLUB	...	... ,,		8 55	...	12 55	1 55	...	5 25	7 25	x 9 35
ASCOT STATION (S.R.)	...	... ,,		9 3	a 10 32	1 3	2 3	a 3 32	5 33	7 33	x 9 43
ASCOT (Horse & Groom)	...	... ,,		9 6	a 10 35		2 6	a 3 35	5 36	7 36	x 9 46
ROYAL HOTEL	...	... ,,		9 8	a 10 37		2 8	a 3 37	5 38	7 38	x 9 48
ROYAL HUNT	...	... ,,		9 13	a 10 42		2 13	a 3 42	5 43	7 43	x 9 53
BROOKSIDE	...	... ,,		9 17	a 10 46		2 17	a 3 46	5 47	7 47	x 9 57
LOVEL HILL (Fleur de Lys)	...	... ,,		9 19	a 10 48		2 19	a 3 48	5 49	7 49	x 9 59
WINKFIELD (Squirrel)	...	... ,,		9 24	a 10 53		2 24	a 3 53	5 54	7 54	x 10 4
WINDSOR HOSPITAL	...	... ,,		9 33	a 11 2		2 33	a 4 2	6 3	8 3	x 10 13
WINDSOR CASTLE	...	... arr.		9 38	a 11 7		2 38	a 4 7	6 8	8 8	x 10 18

From WINDSOR CASTLE to SUNNINGHILL (New Inn).

				a.m.	a.m.	p.m.	p.m.	p.m.	p.m.	p.m.	p.m.
WINDSOR CASTLE	...	... dep.		9 50	11 15		a 2 45	4 15	6 15	8 40	x 10 25
WINDSOR HOSPITAL	...	... ,,		9 55	11 20		a 2 50	4 20	a 6 20	8 45	x 10 30
WINKFIELD (Squirrel)	...	... ,,		10 4	11 29		a 2 59	4 29	a 6 29	8 54	x 10 39
LOVEL HILL (Fleur de Lys)	...	... ,,		10 9	11 34		a 3 4	4 34	a 6 34	8 59	x 10 44
BROOKSIDE	...	... ,,		10 11	11 36		a 3 6	4 36	a 6 36	9 1	x 10 46
ROYAL HUNT	...	... ,,		10 15	11 40		a 3 10	4 40	a 6 40	9 5	x 10 50
ROYAL HOTEL	...	... ,,		10 20	11 45		a 3 15	4 45	a 6 45	9 10	x 10 55
ASCOT (Horse & Groom)	...	... ,,		10 22	11 47		a 3 17	4 47	a 6 47	9 12	x 10 57
ASCOT STATION (S.R.)	...	... ,,		10 25	11 50	1 25	a 3 20	4 50	a 6 50	9 15	x 11 0
MARIE LOUISE CLUB	...	... ,,		...	11 58	1 33	...	4 58	a 6 58	9 23	x 11 8
SUNNINGHILL (New Inn)	...	... arr.		...	12 3	1 38	...	5 3	a 7 3	9 28	x 11 13

a—via Cranbourne Church.   x—Wednesdays, Saturdays, and Sundays.

THE ASCOT PRINTING WORKS.   *Proprietors—C. E. & V. M. JEATT.*

Evidently the additional service was well received, and from 1st July 1932 the original route gained some additional mid-day short-workings between the New Inn at Sunninghill High Street and Ascot Station in place of the previous lay-over time in the hope of an increase in takings or perhaps after local suggestions?

In the meantime a number of issues had arisen over garaging for the buses. Firstly, Beatrice Jeatt, who was not present and had no involvement in the new business, defaulted on the mortgage for Forge Cottage, which put Will and son and daughter Cecil and Vivien out of the property, the effects of which were covered by Hilda Cox's account earlier on.

Here we see Hilda, now Mrs. Nugent, with young daughter Peggy and Will Jeatt outside her cottage around 1934, the Reo (FG 3768) ready for a Sunninghill journey via Cheapside.

As far as the buses were concerned, a rented garage was arranged nearby at 'Meadowscroft' in North Street at a weekly rate of 5 shillings payable to the tenant Sydney Wise. That had started in February 1932, but a disagreement arose in September that year over ownership of some items such as ladders stored there, and Will found the lock had been changed, so he appealed to Mrs. Goddard the landlady.

However, whilst this was going on, negotiations were in hand on a vacant plot of land immediately to the right of Forge Cottage, still in fact the garage site used by *White Bus Services*. It belonged to a retired motor body and coachbuilder William Willis of 32 Grove Road in Windsor, who was willing to lease it at £26

per annum. The plot was 54ft at the road frontage and tapered to 58ft wide at the end of the 175ft depth. As Will was of course an un-discharged bankrupt, the paperwork was made out to Cecil and Vivien, though ironically he was listed as their guarantor. The lease was initially for 7 years, with the option to purchase the land in 1939, a matter actually taken up in 1945 when the war had ended. Conditions imposed meant that all fences and hedges must be maintained, whilst a quince tree was not to be removed.

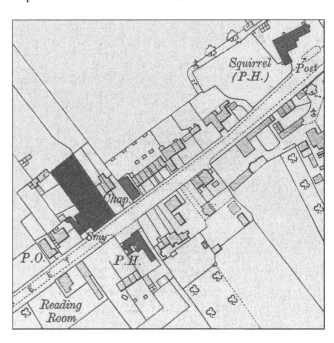

This map extract can still be used today to find the White Bus yard in North Street, with The Squirrel, the former Herne's Oak and the Chapel all highlighted. Also show is the plot of land referred to above, whilst next door and marked Smithy are Forge Cottage and the former North Street Garage.

Although from the day-to-day point of view *White Bus Services* was in the care of C. E. & V.M. Jeatt, sometimes taken by the Traffic Commissioners and others as a husband-and-wife partnership, their father was writing letters on an almost daily basis, to the TC, the *'Valley'* or vehicle suppliers, as well as handling correspondence with Local Councils and enquiries for additional journeys or private hire. Having read all of these communications for the 1930's it is clear that Will had a direct style of writing, but also a good grip on situations, along with a competitive eye.

He continued to berate the unreliability of *Vimmy*, in particular to the missed late-night journeys, which when considered can only have been observed by his presence, presumably in a bus ready to take the poor unfortunates home. This he recorded as doing for free, though no doubt the tired dancers passed the hat about in thanks for not having a tedious walk home in the dark! Both Cecil and his father held licenses as both Driver and Conductor, and in October 1931 daughter Vivien also passed for both.

On the latter point, it has sometimes been noted that she was 'the only lady bus driver in Berkshire' at that time, though there were several others towards the west of the County. She was, however, not the only one to be involved in a local family bus operation, as we have Maud Furlong of Maidenhead and the *Yellow Bus Service* between Maidenhead and Bracknell via Holyport in the 1920's, whilst in the late 1940's we have Vera Gough of *Gough's Garage Bus Service* out from Bracknell to parts of Winkfield Parish and also out to Pinewood and Crowthorne.

1933 – A Dennis Once Again

Both the Cranbourne Institute (built as the Forester's Hall in 1912) and The Cordes Hall in Sunninghill had regular events requiring buses after the last scheduled journeys, and Will Jeatt naturally believed all these should be covered by *White Bus*. Whereas they had them from the Cordes Hall and to Sunninghill from Cranbourne, those to Windsor or the Fernbank Road area had been granted to Bruce Argrave *(Vimmy Bus Service)*. Some of these had been authorised on an ad hoc basis, but the Commissioner granted such journeys over the authorised roads from 23rd July 1933. Again, it must be appreciated that any such event was very popular in those days of no television, as was also the cinema. Indeed, on the latter form of entertainment, escaping to the wonders of the big screen was a great treat, and Windsor cinemas had their posters displayed right by the North Street yard.

Another view of the Reo (FG 3768) outside Wood's the Chemist in the High Street at Windsor. Compare the body on this with the Vimmy Bus Reo on page 15.

During early 1933 a quantity of wood was delivered to the North Street yard, with which a shed was built for the buses, and the completion of that was notified to the agent for the owner on 11th April. However, in due course the Easthampstead Rural District Council noticed the existence of the shed, noting that planning permission had not been sought. In his response Will stated that it was only a temporary building which could be dismantled and that was accepted!

Although generally relations with the *Thames Valley* company were on good terms, Will did write to the Traffic Commissioners stating his surprise that the operator had decided to start a new Route 2c Windsor (Castle) – Winkfield (Squirrel) – Lovel Hill (Fleur-de-Lis) – Brookside (Hatchet Farm) – Shepherd White's Corner – Royal Hotel – Ascot (Horse & Groom) – Sunninghill (Schools) – Sunningdale (Station). In fact the route as far as Ascot followed that inherited from the *Great Western Railway*, whilst the onward link to Sunningdale Station was influenced by the *Southern Railway*, both then shareholders in the 'Valley. The Tilling-Stevens bus inter-worked with Route 2a Windsor – Winkfield – Bracknell in competition with *Vimmy*. The above service commenced on 27th May 1933 with the Summer schedules, whilst dropped from that day was the section onwards to Bagshot, and Will pointed that out to the Commissioner in a manner hinting that *White Bus* might consider filling that void. On the following day he called into see the Commissioner at the SETA office at Somerset House, Blagrave Street, Reading, before popping into *Thames Valley* at 83 Lower Thorn Street to see what small buses they had for sale!

Indeed, he had been looking in May 1933 at a new 20-seater Commer that REAL Carriages had for sale for £500 in Ealing, but wanted them to take the Chevrolet and Reo in part-exchange.

On 6th July he went to the 'Valley's garage at Bridge Street in Maidenhead to view a trio of 1931 Dodge 14-seaters taken over from *Reliance Bus Service* of Slough, a Dennis 30cwt 14-seater (RX 3131) of 1930 taken over from Fred Cowley (*Lower Road BS*) of Cookham, as well as various non-running Chevrolets cleared out from Jimmy Harris (*Pixey BS*) of nearby Fifield. He offered £5 for one of the latter, which was just a chassis for spare parts, £100 for the Dennis and £50 for one of the Dodges, all of which TV turned down. In correspondence Will pointed out that with the recent takeovers under the London Transport Act the market was awash with 14-seaters but to no avail. Although TV realised around £70 each for the Dodges and £9 for the Chevies it got only £30 for the Dennis.

When Cecil was driving for Fred Brown a 30-cwt Dennis bus had been in use, though it didn't pass to the Jeatts. It would seem that it had left a favourable impression though, as in July 1933 a newer replacement G-type model with 17.9hp engine and 11ft 10ins wheelbase came into the fleet, again via E. Blaxton Wilson the dealer. It was new to Sherratt, Booth & Brown (*Invincible Bus Service)* of Biddulph Moor (RF 4355) with 18-seater body of unknown make, and before entering service in September it was thoroughly overhauled at Winkfield, then sign-written by A.A. Best of 'Stonylands' in Egham for 14 shillings and 6 pence, which may have also included the sign erected by the yard entrance around that time.

The Dennis G-type (RF 4355) is seen newly painted in chocolate-and-white but before the sign-writing had been applied, and the sole passenger seems to be Will!

According to the correspondence there were several buses also being prepared by Will Jeatt on behalf of G. Jackson of Dorney Reach near Taplow, certainly a Reo and a Dennis are mentioned during August and September 1933, but fuller details are not known. Also to note is that at that time Will renewed his subs as a member of the Institute of British Engineers.

In a further attempt to increase the late-night trade from the Cranbourne Institute *White Bus* wrote to the organisers in October 1933, offering to put posters up at various points along the service routes for them, as well as reminding them of the open license to operate. He also obtained their support in his complaints regarding Argrave's missed journeys, along with the Police Constable from Cranbourne Police Station, both mentioned in his letters to the Commissioners.

Several outgoings are notable for the year of 1933, one being the 'cab tolls' payable to the *Southern Railway* for use of Ascot Station forecourt, amounting to 5 shillings per week, whilst Ashby's Brewery at Staines charged one shilling a year for the timetable case fixed to the wall of the Royal Hunt in Fernbank Road, North Ascot.

This rear view of Reo (FG 3768) was taken when it was decorated for the 1935 Silver Jubilee, showing Cecil on the rear entrance with pole, with the typical oval rear window of that era. Note also the fleet name contained in a circle, which was presumably applied to the other buses in use at that time.

It seems that the idea of disposing of the Windsor – Winkfield - Ascot – Sunninghill route to *Thames Valley* was considered at the end of 1933, as Will sent the Company details of passengers carried, which showed that in 1932 a total of 43,181 were carried, whilst 1933 had risen to 56,204, accompanied by his seasonal greetings! However the *'Valley'* had a very busy year with takeovers, so no offer was received.

1934 – The First Of The 'Aces'

The Marie Louise Club for Gentlewomen had been opened in 1926 by the Friends of The Poor as its first residential home in South Ascot, and over the years *White Bus* provided various links to and from the local churches for both Catholic and Anglican services on Sundays. However, in January 1934 they wrote to the Club Secretary offering to run journeys to and from the Sunninghill Picture House in the High Street, with return journeys pre-booked for the end of screenings, any such travel being at the normal fares.

Figures for the Windsor – Ascot – Sunninghill route show that in March 1934 passengers carried were at 4907, compared with the same month in the previous year of 4009, around a 20% increase, whilst the income had risen from £49 10s 6d to £61 19s 3s, the latter amount equating to some £3900 value today. In part this increase was probably down to Argrave once again turning his focus to serving the back-roads to Bracknell, details of which will be found on page 16.

March also saw the departure of the little Chevrolet bus (PW 9949), which had not proven the best of purchases despite abundant spares being available for the L-series. That left the 16-seater Reo (FG 3768) and the 18-seater Dennis (RF 4355) covering the bus services and other bookings.

As the Dennis had proven itself to be a reliable vehicle, plus also the fact that the Dennis Bros. works at Woodbridge Road in Guildford, Surrey was only around 20 miles away, that make was chosen for the first new purchase by the Jeatt family. The 'Ace' had a Dennis 24.8hp 4-cylinder petrol engine and a short 9ft 6ins wheelbase, which gave a good turning circle ideal for restricted turns. The front axle was set back from the radiator, which gave the distinctive 'snout' appearance earning it the nickname 'flying pig'.

At that time Dennis also built most its own bodywork at its extensive works at Woodbridge Road, Guildford but in this case the project went to E.D. Abbott over at Farnham, Surrey, though it is not known if by choice of *White Bus* or sub-contracted by Dennis Bros. What is clear is that the little G-type had proven itself a reliable vehicle. When spares were needed they were sent over by passenger train, or as orders sometimes indicated 'will send my girl over tomorrow to collect', a trip out for Vivie!

Unlike the 'snaps' of the White Buses taken by the family, this view of the Dennis 'Ace' was specifically posed for a commercial photographer, outside the Brook Farm Dairy shop by Sunninghill School. Here we see Cecil Jeatt at the wheel and Maurice Nugent as Conductor. Incidentally the centre rear emergency door came in useful if a cycle needed to be carried!

Details of the livery are confirmed by the order for chocolate paint, along with East Kent white and gloss black for the mudguards. As new the bus had white window surrounds, but a later repaint saw them brown instead. As was common at the time, the vehicle was obtained on a hire-purpose agreement through Dennis Contracts based at a London address.

In those days any local events were well supported as long as the sun shone, and a fund-raising fete by the British Legion on 4th August 1934 was covered by a shuttle service of buses between The Squirrel and the venue at Winkfield Place, about 0.75 miles along the Drift Road from the nearby crossroads.

With the anticipated delivery of the new 20-seater, an application was sent to the Traffic Commissioner to extend the Windsor – Cheapside – Sunninghill route onto Windlesham and Bagshot, which would replace the lost *Thames Valley* link between those terminal points discontinued in May of the previous year. Indeed, the Bagshot RDC wrote asking if the extension could be provided. This was the first application actually made under the title 'C.E. & V.M. Jeatt (t / a White Bus Services)', though that did lead to letters addressed to 'Mr. and Mrs.'!

There are no details of fare-taking arrangements up to this point, but in September 1934 an agreement was signed with the Bell Punch Company of St. James's Street, London W1 for 2 Alarm & Registry Ticket Punches, for a period of 5 years at 3 shillings 9 pence per quarter for the pair. Pre-printed values of Bell Punch tickets were used for many years, though in due course a couple of other denominations were printed locally and numbered in the office with a hand stamp. Examples of these tickets will be found on page 113.

The 'Ace' was delivered and registered as JB 4838 on 15th September, and being the first new bus purchased it was very much pride-of-the-fleet.

This offside view of Abbott-bodied Dennis 'Ace' (JB 4838) was actually taken a few years after delivery, by which time the window surrounds were now painted as brown in place of the original white. It is seen near The Crispin emerging from Lovel Lane.

29

White Bus Services. Time Table

From WINDSOR CASTLE to BAGSHOT, via Great Park, Cheapside and Sunninghill

		a.m.	a.m.	p.m.	p.m.	p.m.	p.m.	p.m.	p.m.
WINDSOR CASTLE	dep.	a8 0	a10 0	12 0	2 0	4 0	6 0	8 0	10 0
THE FOUNTAIN	,,	a8 5	a10 5	12 5	2 5	4 5	6 5	8 5	10 5
QUEEN ANNE'S GATE	,,	a8 7	a10 7	12 7	2 7	4 7	6 7	8 7	10 7
FOREST GATE	,,	a8 14	a10 14	12 14	2 14	4 14	6 14	8 14	10 14
CRISPIN HOTEL	,,	a8 17	a10 17	12 17	2 17	4 17	6 17	8 17	10 17
WOODSIDE ("Duke of Edinburgh")	,,	a8 19	a10 19	12 19	b2 19	4 19	6 19	8 19	c10 19
SUNNINGHILL PARK	,,	a8 21	a10 21	12 21	b2 21	4 21	6 21	8 21	c12 21
SAWYER'S GATE	,,	a8 24	a10 24	12 24	b2 24	4 24	6 24	8 24	c10 24
CHEAPSIDE	,,	a8 28	a10 28	12 28	b2 28	4 28	6 28	8 28	c10 28
BLACKNEST	,,	a8 33	a10 33	12 33	b2 33	4 33	6 33	8 33	c10 33
SUNNINGHILL ("Cannon")	,,	a8 37	a10 37	12 37	b2 37	4 37	6 37	8 37	c10 37
SUNNINGHILL	,,	a8 40	a10 40	12 40	b2 40	4 40	6 40	8 40	c10 40
BERYSTEDE HOTEL	,,	a8 42	a10 42	12 42	b2 42	4 42	6 42	8 42	c10 42
SWINLEY TURN	,,	a8 45	a10 45	12 45	b2 45	4 45	6 45	8 45	c10 45
THE "WINDMILL"	,,	a8 48	a10 48	12 48	b2 48	4 48	6 48	8 48	c10 48
BAGSHOT (The Square)	arr.	a8 54	a10 54	12 54	b2 54	4 54	6 54	8 54	c10 54

From BAGSHOT to WINDSOR CASTLE, via Sunninghill, Cheapside and Great Park

		a.m.	a.m.	a.m.	p.m.	p.m.	p.m.	p.m.	p.m.	p.m.
BAGSHOT (The Square)	dep.		a9 0	a11 0	1 0	b3 0	5 0	7 0	9 0	c11 0
THE "WINDMILL"	,,		a9 6	a11 6	1 6	b3 6	5 6	7 6	9 6	c11 6
SWINLEY TURN	,,		a9 9	a11 9	1 9	b3 9	5 9	7 9	9 9	c11 9
BERYSTEDE HOTEL	,,		a9 12	a11 12	1 12	b3 12	5 12	7 12	9 12	c11 12
SUNNINGHILL	,,		a9 14	a11 14	1 14	b3 14	5 14	7 14	9 14	c11 14
SUNNINGHILL ("Cannon")	,,		a9 17	a11 17	1 17	b3 17	5 17	7 17	9 17	c11 17
BLACKNEST	,,		a9 21	a11 21	1 21	b3 21	5 21	7 21	9 21	c11 21
CHEAPSIDE	,,		a9 26	a11 26	1 26	b3 26	5 26	7 26	9 26	c11 26
SAWYER'S GATE	,,		a9 30	a11 30	1 30	b3 30	5 30	7 30	9 30	c11 30
SUNNINGHILL PARK	,,		a9 33	a11 33	1 33	b3 33	5 33	7 33	9 33	c11 33
WOODSIDE ("Duke of Edinburgh")	,,		a9 35	a11 35	1 35	b3 35	5 35	7 35	9 35	c11 35
CRISPIN HOTEL	,,	a7 38	a9 38	11 38	1 38	3 38	5 38	7 38	9 38	c11 38
FOREST GATE	,,	a7 40	a9 40	11 40	1 40	3 40	5 40	7 40	9 40	
QUEEN ANNE'S GATE	,,	a7 47	a9 47	11 47	1 47	3 47	5 47	7 47	9 47	
THE FOUNTAIN	,,	a7 49	a9 49	11 49	1 49	3 49	5 49	7 49	9 49	
WINDSOR CASTLE	arr.	a7 54	a9 54	11 54	1 54	3 54	5 54	7 54	9 54	

a—Not Sundays. b—Fridays, Saturdays and Sundays. c—Saturdays and Sundays.
Additional Departures from Bagshot leave five minutes after the hour.

Oxley & Son (Windsor) Ltd., 4 High Street, Windsor. Proprietors—C. E. & V. M. JEATT.

The timetable sheet for the Windsor – The Crispin – Sunninghill – Bagshot service as from July 1935.

The application for the new extended route to Bagshot was actually quite sparse, with only 3 through journeys between Bagshot and Windsor each day other than Saturdays, when the frequency was effectively improved from 4 to 2-hourly, and several other journeys were only short-workings between Windsor and The Crispin. However, between placing the application and approval the service over the 14-mile route was much enhanced as shown above.

The new license was approved on 1st July 1934, but not issued until 15th October, so it has not proven possible to definitely ascertain the start date of the extended route, also bearing in mind that the new bus arrived on 15th September, the generally accepted month of change.

1935 – The Silver Jubilee Year

From January 1935 an agreement was made with J. Purchase of 230 Acton Lane, London W4 to provide printed advertisements for local shops etc., each being displayed inside the buses at 7 shillings 6 pence per annum, which was a useful income to offset other costs. The number of passengers carried on the route via Ascot that March showed only a small increase on the previous equivalent figure by only 17, though the income was improved at £78 9s 5.5d, whilst the route to Bagshot brought in £57 10s 1.5d that month with 2903 passenger journeys.

However, the real theme for 1935 was the Silver Jubilee of the reign of King George V, the official date of 6th May being set, though associated events were wider spread. Indeed, one very popular excursion that Summer was to see the Royal Navy ships assembled off Spithead for the Royal Review, and many enjoyed the trip to the coast, with a number of firms staying into the evening to see the illuminated vessels.

More locally there were fetes and parades, what with Windsor being a Royal town, so it was a busy year at *White Bus*. The King, along with Queen Mary were popular monarchs, and the Company entered into the spirit of the celebrations by decking out the fleet with flags and bunting.

In the light of experience the 2.28pm departure from Cheapside to Bagshot was reduced in frequency to Fridays, Saturdays and Sundays only, whilst the 10pm from Windsor was altered to run through to Bagshot at 11pm and return to The Crispin at 11.38pm just on Saturdays and Sundays, both of those changes being reflected in the July 1935 timetable as shown above.

Approval was also given from July for special later journeys, operated at least 30 minutes after the last scheduled buses, to take passengers from the Cordes Hall in Kings Road, or the Comrades Club in Bagshot Road, both Sunninghill to any point along the route towards Sunningdale and Bagshot, or towards Ascot, Cheapside and Winkfield. At the same time a link for similar events was approved between the Royal Albert Institute in Sheet Street, Windsor and Wheeler's Yard or Cumberland Lodge, both the latter within the private road network of Windsor Great Park and for the benefit of residents there.

'Ace' JB 4838 is seen at the North Street yard having been decorated for the 1935 Jubilee, though in fact the Road Traffic Act specifically forbade the trailing of flags and other items from PSV's, a matter no doubt overlooked for the national celebrations.

One of the sensations of that age was 'Sir Alan Cobham's Flying Circus', actually touring for the last time that year. The famous founder had been a pilot in the Royal Flying Corps during the Great War, after that becoming test pilot for the de Havilland aircraft company. He undertook a 5000-mile tour of Europe in 1921, then in 1926 flew a de Havilland DH60 float-plane to Australia and back. In 1932 he established the National Flying Days in order to stimulate the public interest in the opportunities flight could bring.

Such an event was understandably very popular, so when it came to the Winkfield Flying Ground over at Foliejon Park on Winkfield Plain it drew large crowds including those travelling by bus. To cater for the day on Friday 27th September, *White Bus* laid on a special shuttle service from Crouch Lane and The Squirrel to the ground via the Drift Road and Maidens Green, a distance of some 1.5 miles at a fare of 3d per journey.

October saw another round of enquiries regarding the possible purchase of a small Dennis, with one at *Premier Coaches* of Watford and others from dealer Arlington Motors being considered but not purchased.

The figures for the full year show that 42,131 had used the Windsor to Bagshot service during 1935, but no corresponding information is available for usage on the original Windsor – Ascot – Sunninghill route, though *Thames Valley* was now reviewing that area.

1936 – A Second 'Ace' Arrives

This year would see a number of changes to the bus operations of *White Bus Services,* and at this point it is worth remembering that although it ran into Windsor, which was within the designated area under the London Transport Bill of 1933, its Winkfield base and routes further afield had saved it from compulsory purchase as had been the fate of so many local bus firms with origins in the 1920's. Indeed, even *Thames Valley* had to surrender its Windsor – Staines service, along with a pair of double-deckers.

However, the latter had also been looking at eliminating the competition on the Windsor – Ascot – Sunninghill road, so it made approaches in the early months of 1936 to both the Jeatts and the Argraves over their respective operations along that road, and terms were drawn up for each during March 1936.

It has been reported previously that in fact *White Bus* never signed the agreement concerning the takeover, though fuller investigation reveals that there is more to just the surviving un-signed copy of the agreement, which is actually the original version sent by *Thames Valley.* Within that the right to use the name *White Bus Services* is removed, as was the opportunity to start other services within the area.

Will Jeatt objected to those provisions and a further version was drafted which met with his approval, and behind the scenes he elicited an understanding that once *Vimmy* was acquired, the right to run the late-night buses from the Cranbourne Institute towards Windsor would be transferred to him, which was agreed to by Traffic Manager John Dally, with a period of 7 years free of objections by the *'Valley* also forthcoming, as well as not objecting if the Bagshot service was to be operated more frequently.

Thames Valley paid £1,250 for the license for the Windsor – Ascot – Sunninghill service, but no vehicles were involved, the transfer being on 13th May. It was indeed a good return on the original investment when the route and bus were purchased from Ackroyd, and there were already plans for using this money to further expand operations, and also to add a second new Dennis 'Ace' to the fleet.

Generally the first 'Ace' had proven a very reliable type, though there were issues with the steering which required frequent attention, and at this time Dennis had suggested trying some different tyres. The second example was actually ordered through Abbotts, who had a showroom in Bagshot which Will called into to discuss it and the possible purchase of a Dodge or Bedford to be part-exchanged for the Dennis G (RF 4355) and the Reo (FG 3768), this being the first mention of the Bedford make that would feature much in the fleet in due course! Other enquiries in the

Spring were made to Willmotts Motors of Hammersmith for a 1933 Bedford and to *Southdown Motor Services* of Brighton for a Dennis G-type.

The first 'Ace' featured a Smith's Clock in the front bulkhead, which was repeated for the new order, but also a Philco Radio was ordered at a extra £18 plus £7 for fitting, making it a popular bus to ride on, a good vehicle for private hire and a nice backdrop for the driver, radio being the main form of entertainment. The bus would again be purchased on hire-purchase through Dennis Contracts, though the opportunity of the new money recently injected allowed for the outstanding debt on the first 'Ace' to be cleared. Also with this example the body would be by Dennis at Guildford, though in due course they did send it to Farnham for Abbotts to paint during June 1936.

The second 'Ace' (JB 9468) was the most widely photographed of all the earlier fleet, and is seen here outside a pub, which may well be either for a dart's match or one of the regular outings so popular then.

Also forthcoming during the early months of 1936 were representations of behalf of residents of the Park to have the Windsor – Bagshot service diverted to serve their community, and as these were headed by Eric (later Sir Eric) Savill, we will take a look at his background and influence.

Educated at Malvern then to Magdalene College, Cambridge, his studies were interrupted by service in the Great War, where he sustained injuries in 1916 at the Battle of The Somme. After completing his studies he joined his father's successful London practice of Land Agents and Chartered Surveyors. However, an old friend was duly appointed as Librarian at Windsor Castle, leading to frequent visits to the area by Eric, who got to know the Great Park very well. Then, quite to his surprise in 1931 he was offered the post of Deputy Surveyor of Parks & Woods by the Crown Commissioners, a post he accepted despite his future assured in a lucrative partnership.

Savill became Deputy Ranger of the Great Park in 1937, by which time he had already earned the respect of the monarchy for his efficient yet often innovative management of the 15,000 acres under his control, and in the early 1930's he started the project of the magnificent gardens that now bear his name.

However, he not only addressed the needs of the land, but also did much to improve the lot of those who were employed in the Park, as well as for their families, with a programme of cottage building, better social facilities and, as now revealed by research, a bus service! In respect of accommodation there were cottages built in 1934-8, with 8 at Mezel Hill and 2 at Cumberland Gate and Queen Anne's Gate, as well as other groups outside the Park boundary. A 'model village' was started in 1948 with a Post Office Stores and 32 cottages there, followed by 18 more in 1954 and a further 6 in 1966, and all of these developments have a bearing on future *White Bus* operations.

Savill's personal involvement in bringing the bus service into the Park becomes evident through the surviving correspondence, whilst it also answers the question no other researchers seem to have thought to ask as to how the Winkfield-based operator came to be running over the private roads, quite simply it was invited!

In request to the initial suggestion of diverting the Windsor – Bagshot route, Will Jeatt responded that it would not be practical, but that he had applied for a new service to start from The Crispin Hotel, not far from Forest Gate. The bus would enter the Park there, then pass the Prince Consort Workshops (near where The Village would later be developed), then turn away from the Windsor direction to Cumberland Lodge, before continuing a circular route passing Chaplains Lodge (near Bishops Gate), the Copper Horse, Beech Hill Lodge and out at Ranger's Gate to emerge on the public highway of Sheet Street Road into Windsor and terminate at the Castle.

The service was intended to cater mainly for shopping needs, bearing in mind that domestic refrigerators were a rarity then, so the purchase of perishable goods was spread over the week. On the original application only Tuesdays and Fridays were to be covered, but a further letter from Eric Savill saw the addition of the same 2 return journeys on Saturdays as well, with the benefit also extended to the leisure of those working during the week. The license was granted from 1st March 1936, beginning the service to the Park by *White Bus* which has continued in its various forms through to the present day.

The fares on the new service for the main points were: Windsor to Queen Anne's Gate 2d single, no return; Windsor to Cumberland Lodge 7d single, 10d return; Windsor to the Crispin Hotel 6d single, 10d return. Contained in the conditions of the license was that no double-deck vehicles could be used, and the speed limit in the Park of 38mph must be observed.

WHITE BUS SERVICES. Time Table.

TUESDAYS, FRIDAYS AND SATURDAYS

CRISPIN HOTEL TO WINDSOR CASTLE
via CUMBERLAND LODGE.

				p.m.		p.m.
CRISPIN HOTEL	...	...	dep.	2 19		6 19
FOREST GATE	...	...	,,	2 21		6 21
ISLE OF WIGHT POND	...	...	,,	2 23		6 23
CUMBERLAND LODGE	...	...	,,	2 30		6 30
CHAPEL LODGE	...	...	,,	2 34		6 34
COPPER HORSE	...	...	,,	2 39		6 39
BEECH HILL LODGE	...	...	,,	2 44		6 44
QUEEN ANNE'S GATE	...	...	,,	2 47		6 47
THE FOUNTAIN	...	...	,,	2 49		6 49
WINDSOR CASTLE	...	...	arr.	2 54		6 54

WINDSOR CASTLE TO CRISPIN HOTEL

				p.m.		p.m.
WINDSOR CASTLE	...	...	dep.	3 0		7 0
THE FOUNTAIN	...	...	,,	3 5		7 5
QUEEN ANNE'S GATE	...	...	,,	3 7		7 7
BEECH HILL LODGE	...	...	,,	3 10		7 10
COPPER HORSE	...	...	,,	3 15		7 15
CHAPEL LODGE	...	...	,,	3 20		7 20
CUMBERLAND LODGE	...	...	,,	3 24		7 24
ISLE OF WIGHT POND	...	...	,,	3 31		7 31
FOREST GATE	...	...	,,	3 33		7 33
CRISPIN HOTEL	...	...	arr.	3 35		7 35

Luff & Sons Ltd., Printers, 47, St. Leonard's Road, Windsor.

The timetable of the Windsor Castle – Crispin Hotel service shown half-size. Note that the columns are set with blanks between them in order to make it cheaper should extra services be added later, all such setting being by hand using individual type characters then.

Dennis 'Ace' (JB 9468) is seen outside the Crispin Hotel during 1937, when it would be decorated for the Coronation. The sign to the right includes a board stating 'No charabancs except by appointment' even though the new larger establishment was built to cater specifically for passing trade in the motor age.

The license for the new service was effective from 1st March 1936, and on 28th of that month an order was placed with Luff & Sons Ltd., stationer's and printers of St. Leonards Road, Windsor for 5000 flyers for the Bagshot service and 3000 for the Crispin route. Also in connection with the changes, an order was placed at the end of May with Bell Punch for 2 Ticket Boxes, along with an initial batch of 10,000 tickets.

As to the start date of the new service there is again no firm evidence, the Road Service License being issued on 10th July 1936, but in the local travel guide for July, and the new 'Ace' was collected from the works on 2nd July and registered JB 9468 on that day. Preparations had already been made to cater for the other popular event of June, the Aldershot Military

Tattoo, with *White Bus* gaining a license to run to that annual event, which that year saw rehearsals on 11-13th and the main event on 16-20th, the latter coinciding with Royal Ascot Race Week. Approval was given for pick-ups in North Street, any points in the Park or along the Bagshot route as served by the buses, with fares of 3s 6d for the daylight rehearsals, 6s for the Saturday night finale and 4s 6d for the other evening performances.

There are few details of the private hire work undertaken in those earlier years, but the cost of hiring a 20-seater for a trip to Brighton or Worthing was then £5 5 shillings or to Southend for £6.

As soon as the deal with *Thames Valley* over the service via Ascot had gone through Will called into the Ascot Station to inform the Station Master of the change, and also of course to cancel the cab tolls. He did, however, object to the *'Valley'* when he saw its flyer announcing it 'had absorbed *White Bus Services'* so he wrote to Mr. Dally pointing out it had cost him private hire work and demanding its withdrawal.

As it was John Dally had quite a bit on his mind that June, as the crews of that Company held a strike from 14th June to 2nd July, now known to be a politically-motivated attack by the Union on the Tilling Group, of which *Thames Valley* was a member at that time.

As a result of the above situation Will Jeatt wrote to the Traffic Commissioner offering to run a limited service along his former route between Windsor and Ascot until the strike had ended, adding that 'we are on good terms with the Company and its men'. There are no records of this actually running, but even the Ascot-based crews would not have really objected as their families still needed to get about!

One of the many posed photos from the family album of Dennis 'Ace' (JB 9468) with one of the ancient oak trees near the Stone Bridge and The Copper Horse. Some of the trees are over 600 years old!

On 7th July 1936 The King reviewed the Life Guards on the Review Ground off Sheet Street Road, so an enhanced service was provided on the Bagshot route, whilst due to the uncertainty caused by the above strike, permission was also obtained to run a special journey from the Marie Louise Club to the event.

The search had continued for another good used 20-seater bus, and on 18th August a Commer 'Centaur' T20X-type (JB 568) was placed in service, having been acquired from G.J. Dawson, the dealer in Wimbledon, London SW9. The bus had started life in Windsor with Martin & King (*Nippy Bus Service)* in 1932, but passed to *London Transport* when that operator was taken over in February 1934 under the 1933 Act, seeing service with them until early 1936.

The Commer 'Centaur' (JB 568) was a nice bus with a 20-seater front-entrance body of unknown make but with a distinctive hump to the roof. It is seen outside the Hernes Oak pub opposite the North Street yard, freshly repainted in brown, white and black.

In order to complete the equipment for the new 'Ace' and the Commer Will ordered destination blinds, but in doing so put Crispen on the order, which for some time afterwards was displayed! As to how the fleet was distinguished, now that all 3 bore the 'JB' mark, there are some references to 'Ace 1' and 'Ace 2', whilst 'the Commer' was sufficient, though later practice was to use the registration letters of vehicles.

Also earmarked from Dawson was a Dennis which had been stated as a late 1933 GL-type (KX 8569) which Will hoped to part-exchange the Dennis G-type (RF 4355) and Reo (FG 3768), both of which had been withdrawn on the day the Commer arrived.

However, when Will went over to see the bus he was not happy with its condition, the fact that it did not have new tyres, and also that it bore an older-style radiator. He even checked with *London Transport,* as the bus had been new to its subsidiary *Amersham & District* Company until absorbed under the 1933 Act, as well as asking Dennis for confirmation of details. From that found that it was actually built in April

1932 and being an early example bore a G-type radiator, so he refused to accept it. However, the firm would not return his £50 deposit, plus the Dennis and Reo remained in the yard awaiting part-exchange, the matter rumbling on from August to December before recourse to his Solicitor saw Will's deposit refunded. As a postscript to that unfortunate episode, a short time afterwards Dawson's representative handling that was jailed for falsely advertising other vehicles.

Also during August enquiries were made of Willetts, the dealers and coachbuilders of Colchester for a 1932 Bedford, also intended in exchange for the withdrawn buses, though that make would not feature until 1948.

Back in May Vivie Jeatt obtained her first motor car, a 1933 Austin 7.8hp 4-seater saloon (JB 1481), at which point she was living at No.1 Crouch Lane, just off North Street. However, despite that apparent sign of affluence, the services were only marginal in their income, not providing much above costs. One week in August 1936 shows that costs were £20 1s 2d, which included wages for Mr. Jacobs (driver) £3, Cecil at 17s 6d and Vivie at 10s, whilst income from the services was £25 10s 5.5d, plus £5 3s 4d private hire.

Hilda and Peggy Nugent stand with Vivie by her 1933 Austin car (JB 1481), with the withdrawn Dennis and Reo buses standing in the yard at North Street.

During December 1936, and obviously inspired by the potential for passengers during Royal Ascot Race Week, *White Bus Services* applied for an 'express' license to run a direct service between Windsor and the racecourse, which was granted subject to the usual conditions relating to the Police arrangements on authorised routes for traffic management of that very busy event. At the same time permission was given to run a bus around midnight after dances held at the Horse & Groom pub in Ascot High Street throughout the year, the *'Valley* having no objection, as its services ended much earlier.

Despite the obvious energy Will put into managing *White Bus* he was not in good health, and the early part of the year saw him with bronchitis, and when 'laid up' some correspondence was handled by Vivie.

Will Jeatt leans on the gleaming Dennis 'Ace' (JB 9468) in the Summer sunshine of 1936, with the Reo 'Pullman Junior' (FG 3768) and Dennis G-type (RF 4355) withdrawn from service. Note also the pair of cottages behind and the Wesleyan Methodist Chapel, which will continue to feature in later photos. Just to the right are the local cinema bill boards.

1937 – The Coronation Year

There was a postscript to the affair of the Dennis GL (KX 8569), when in January 1937 the County Court judged in Will's favour that any contract over it had been cancelled. During that same very cold month a pair of radiator 'muffs' were obtained from Dennis, and Vivie drove over to pick them up to improve the starting of the buses in the mornings. That month also saw two lorry-loads of brick rubble delivered to make up a hard-standing area.

With the Dawson issue settled there still remained the matter of disposal of the Reo and Dennis buses, and to that end the dealers Worthmores were asked to give details of secondhand stock or what they might pay in part-exchange against a new bus. In the meantime January saw the original 'Ace' being overhauled, the engine having done 125,685 trouble-free miles, followed by attention at Abbotts, which included a new roof (which had leaked) and repaint.

As a result of the death of King George V in January 1936, he had been succeeded by son David as King Edward VIII and a coronation date set for 1937, when in December 1936, the latter abdicated, which

then put King George VI taking up the planned date of 10th May 1937 at Westminster Abbey. Apart from the actual procession and ceremony on the day, the event sparked a similar range of celebrations as the Jubilee of only 2 years before. The Tattoos at Aldershot and Tidworth on Salisbury Plain were both enhanced, whilst another Naval Review was included as well as many local fetes. Such was the demand for travel that the Traffic Commissioners suspended the requirement for applications for enhancements of licenses already held, though Police restrictions had to be observed on routes and parking in the capitol. It is notable that with that event *White Bus* gained private hires from both Bagshot and Chobham.

The White Bus board outside the shop in Bagshot Square in the 1930's, with cat posing in the sunshine.

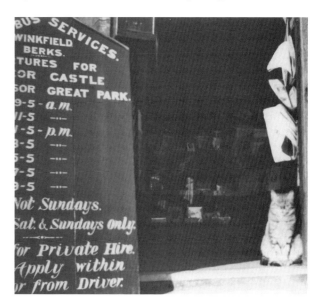

This April 1937 photo by D.W.K. Jones shows the buses in the roadway by The Guildhall at Windsor, the White Bus Dennis 'Ace' (JB 9468) on the service through the Great Park to the Crispen Hotel, whilst Thames Valley's Ascot-based 1927 Tilling-Stevens B9A Car 148 (MO 9320) covers the former route of the Winkfield operator to South Ascot, with a London Transport ST-type about to pass them on its way to the Board's own terminus off St. Leonards Road and opposite the King Edward VII hospital.

A review of some journeys on the Windsor – Bagshot service saw refinements from 1st July, whereby the 10.5pm from Windsor only ran beyond the Crispin for those already on board as far as Sunninghill High Street only, the through journey to Bagshot and back now only operating on Saturdays and Sundays. Other minor changes were made in response to the working times of the Crown Foresters.

In common with most other rural independents, little reference was made to travel in the wider world, so it is notable that *White Bus* publicity included the use of the express coach services of *Thames Valley* (the Reading – London route at The Cannon), also from there *Green Line* (Ascot – London and onward to Gravesend in Kent!), as well as *Aldershot & District* (Farnham – London coach, plus buses to Camberley, Weybridge, Egham and Aldershot via Bagshot). Also noted were the bus services of Harry Lintott of Lightwater *(Direct BS)* from Bagshot – Guildford and Bagshot – Bracknell, the times of which were carried by *White Bus,* and we shall hear more of those soon.

On 23rd October we have the first known accident involving one of the buses, in Sunninghill High Street, when one ran into the stationary Morris van (ABL 921) of the Royal Albert Laundry, and although the damage to the bus was slight, the driver was deemed to have been at fault, so the cost fell on *White Bus.*

1937 had been a busy year all in all, and the Windsor – Bagshot service had carried 69,871 passengers, as opposed to the 42,131 of only 2 years before, with an income of £1373, plus a further £100 by the limited service through the Park and £287 from private hire.

1938 – Bagshot to Bracknell?

There is a noticeable increase in private hire work from 1935, though few details have survived, that due to the improved vehicles now on offer and the general reduction in competition as a result of the take-overs by *Thames Valley* and the London Transport Act. In one amusing reminders of those earlier days, one party organiser sought a 23-seater for a job, to which Will responded that 'as some of them are recalled as rather small one of the 20-seaters would suffice'!

By February 1938 the Commer had been paid off, and in fact no further attempts were made to buy another vehicle that year. When the second 'Ace' was over-hauled in February after 62,000 miles the cylinder block was found to be so clean that Will sent photos to Dennis, used by them in 'The Dennis Distributor', as well as to the oil supplier Redline Glico.

The second Dennis 'Ace' was photographed by the family throughout the Great Park, so these two pages offer views of it at points along the route.

Top - *Here we see the bus at the lodge by Queen Anne's Gate, suitably decorated for the coming coronation as it heads for the Copper Horse – Cumberland Lodge – The Workshops and then The Crispin Hotel. The roadways within the Park were described as early as 1822 by the traveller and writer William Cobbett as well-made, a century prior to the other local roads being surfaced, most being done after the Great War!*

Middle - *Mention has been made of Vivie gaining her PSV license, and this photo shows her at the wheel of the second Dennis 'Ace'. Note the fare table in a frame, display of which was required under the 1930 Road Traffic Act. The ash-tray has been turned up-side-down behind her, presumably to deter use!*

Bottom – *Passing a white-painted cottage on the way into Cheapside, the bus is entering a settlement which predates Sunninghill and Sunningdale by centuries. Various hamlets grew up on the fringes of the Park at the junction of east-west and north-south track-ways, in this case in the triangle created by the modern-day Watersplash Lane, Cheapside Road and Sunninghill Road. When the White Bus arrived in Cheapside it was in fact the first service that area had seen, a continuous one having been operated ever since then, used by many children and others locals.*

Top – Both of the 'Aces' were caught here outside Cranbourne Court, and were probably being used for one of the local events of the coronation, as they carry the decorations of that year. This view also gives a comparison of the Abbott and Dennis bodies, the former with the straight waistrail and the latter with the downward curve to the front dash. Note the man with the hand-mower tidying up the verge and, once again the very large trees right up to the road, a typical situation locally of a road system through the ancient forest area.

Middle – The second 'Ace' passes Forest Lodge, built in 1836, and this shot affords a good view of the road towards Fernhill and Ascot. Note also the actual gate-post with a lamp on top, these being the only lights in the area, so local knowledge was a necessity, whilst the letters on the road warn motorists that they are about to pass over a cattle-grid. The signage points out that the roads are private, along with the Bye-laws relating to the use of the area.

Bottom – The same bus is seen turning at the junction of Watersplash Lane and Sunninghill Road to enter Cheapside, again in 1937. Over the years the timings got quite complicated once the Sunninghill and Park timetables were combined, with lots of variety over the points served, as well as qualified by the days of operation! However, in those days the bus drivers did not have to contend with much other traffic, and certainly not the road-side parking which makes sections of the route tricky.

During March 1938 Will wrote to the Windsor Division of the Berkshire Education Committee to let them know that he had a bus available at times, using its letter informing him of the change of local school times as a lead. From that came a reply that quotations were being sought for transport to swimming lessons at the two local public pools at The Pantiles on the A30 near Bagshot and Noyes Pool, off Broad Lane in Bracknell. These ran during June and July and were part of a national safety campaign to teach children to swim, as drowning was then one of the highest claims on young lives. The quote was successful, so a weekly schedule was drawn up for schools at Cranbourne and Ascot Heath to go to Bracknell, whilst Sunninghill and Sunningdale went to Bagshot, each coach transfer at 10 shillings return for a 20-seater. It is notable that some 77 years later *White Bus* still takes children to swimming, though now of course to heated pools!

The bunting above their heads would indicate that this photo of Cecil and Vivie Jeatt was taken in 1937, the latter wearing her PSV Driver's Badge, the location being the North Street yard opposite the Hernes Oak.

During the Spring of 1938 some interesting events took place which could have seen a further expansion of the *White Bus Services*. It started in the latter part of the previous year when Sidney Ansell, a Peckham-based coach operator acquired the bus and garage business of Harry Lintott *(Direct Bus Services)* based at Lightwater. We have already heard that the latter had a Bagshot – Guildford service, which had started in April 1928, and then in November 1931 he had extended 4 of those journeys in the opposite direction to Bracknell, also running them on Sundays as well.

By late 1937 Lintott was ill and in trouble with the Traffic Commissioner, despite his wife and son trying

to keep things going, so he sold out to Ansell, who was looking for a semi-retirement away from London. However, he duly found he wanted to dispose of the bus services, which he offered for sale in Commercial Motor, which the Jeatts saw and asked for details.

To complicate matters the local territorial company, *Aldershot & District* also took an interest, so we have two strands of correspondence and applications at the same time. In essence the Aldershot company was only interested in the Bagshot – Guildford link, so when it applied to the Commissioner the 'lack of chimney pots over the section to Bracknell' was accepted as being the reason to no longer cover, the decision noting that, 'should in the future a public demand be forthcoming, the Commissioner would expect *A&D* to provide such a service'. It should also be appreciated that the operator had shares held by the *Southern Railway,* which contended that its links provided sufficient means to meet public needs using its trains with a change being required at Ascot.

On the other hand the Jeatts foresaw the opportunity to re-arrange some existing workings to provide a through service as Windsor – Bagshot – Bracknell, and in that respect they had the support of the local Easthampstead Rural District Council.

After the initial letter on 28th February Will took up Ansell's invitation to meet with him at Lightwater, after which the latter confirmed he was offering the licenses for the bus services and excursions and tours at £750, with another £1750 for a Chevrolet LQ-type 14-seater of 1930 (UR 5780), a 1935 Bedford 20-seater (ACG 89), both bodied by Thurgood, as well as a 1936 Reall-bodied Dodge RBF-type (BOR 501). He also had a Dennis 'Pike' coach with Thurgood 20-seater coach body (EXF 377), which he intended to retain at that point in time.

The Direct BS Bedford came close to becoming the first of its make with White Bus, seen in this photo by Joe Higham, whilst the Dennis 'Pike' coach would also have fitted nicely in the fleet with the 'Aces'.

The informative correspondence between Will, Ansell and the *White Bus* advisors reveals that 'the money from the *Thames Valley* purchase is still available', but adding some concern regarding his own health, before going on to say 'would make a nice through service', though again in the way of caution he suggested to his advisor that maybe the *Valley* might like to run over the Bagshot – Bracknell road, as 'they have always worked with us in a friendly manner'. As he was 'laid up these last few weeks, he had then given it more thought and would like to expand'.

In response he was advised to set up a limited company to take over the *White Bus* and *Direct Bus* interests, with the family holding the shares and Sidney Ansell a debenture, as suggested by the Mr. Eustace he was consulting. Will then noted that 'we could start another service through the Park to Virginia Water without upsetting *Thames Valley'*, and adding that *London Transport* ran its Windsor to Guildford service via Staines, which gives further emphasis to the unique setting of the Great Park area.

It should be appreciated that the proposed new route was nothing to do with the existing service to Bagshot from Windsor via Sunninghill, but would incorporate the limited service so far operated through the Great Park, travelling out from Windsor just as already the case through the Park to The Crispin, where it would continue southwards along Lovel Road to Shepherd Whites Corner, where it turned left to the Royal Hotel on the corner of Ascot High Street and Windsor Road, before proceeding straight across what was then a crossroads to Kings Ride. As it descended Kings Ride it passed over the Waterloo to Reading railway, then past the Royal Berkshire Golf Course, followed by Swinley Turn (for Bracknell), and onto the A30 and over Cricketers Bridge into Bagshot Square.

A schedule was put together which shows the bus starting out at The Crispin at 8.33am and arriving at Bagshot at 8.56am, whilst the full journey from Windsor would be 51 minutes. That first bus would then leave Bagshot (The Square) at 9.10am as the run to Bracknell (Hinds Head, top of the old High Street and Cattle Market venue then) via Cricketers Bridge – Swinley Turn – Golf Links – Nine Mile Ride Turn – Bagshot Road (Horse & Groom), arriving at 9.27 to form the 9.30am return trip. After returning to The Square at 9.47 it formed the 10.05am through to Windsor, after which it repeated the full coverage of both routes in similar fashion through to working back short to The Crispin, where it finished service at 10.28pm, giving a 4-hourly headway through the day.

These changes would also see the service through the Park increased from only Tuesdays, Saturdays and Sundays to a Mondays to Saturdays operation. Fares were proposed at Windsor – Bagshot 1s 7d single and 2s 4d return, and Bagshot – Bracknell 8d single or 1s return, and the application was submitted. A map was also included which showed the existing services, as well as those of *Thames Valley,* which were avoided.

The White Bus presence in Bagshot was already well established, and here we see Will Jeatt holding a child, with the first 'Ace' (JB 4838) waiting behind.

In the meantime Ansell had decided to dispose of his service to *Aldershot & District* on 1st June, though it only wished to continue the Bagshot – Guildford link, which at least meant theoretically that the *White Bus* proposals could still be considered as intended in order to improve an existing service, open up a new link to Bagshot and also provide a route to Bracknell, all without competing with other bus services.

In the end the Traffic Commissioners had the hearing at Middlesex Guildhall on 5th October 1938, when the *A&D* assertion that a Bagshot – Bracknell link was not required was upheld (with the proviso of possible future needs), which left *White Bus* no option but to withdraw its other related proposals for the changes and extended route.

The Bagshot – Guildford service became *A&D* Route 62, whilst the road to Bracknell was not actually covered by it until the joint Route 75 with *Thames Valley* commenced in April 1954.

Apparently it had come to the attention of the Traffic Commissioner that the bus on the Park service had been deviating to Bishopsgate, and the firm was warned to keep to the route. It was probably just in order to drop someone off nearer their residence, but was to have been included on the proposed route.

Mention was made earlier of some building work to house the buses at North Street, and indeed a timber shed initially for two buses was erected, but during October 1938 further work took place, allied with the creation of a better forecourt, which was probably when the accommodation reached the 4-bus stage which was evident for some years afterwards.

With apologies for the use of such a poor photo, but this allows us to see what was occurring on the site at North Street, with the old body off the Republic bus to the left and the rear half of the timber bus shed to the right and in the course of construction.

No application was ever made for the proposed route through the Park to Virginia Water, but if that had been extended onto the station bearing that name, which is some 1.75 miles from the lake-side, it would have certainly proven popular with day trippers.

1938 is the first year for which full details of all private hire jobs are available, so we shall a look at the type of work undertaken, along with the hirers. A very regular aspect of work in the evenings was taking local pub dart teams to other venues, all within an orbit of places served by the bus services generally, sometimes with several each evening done in relay. Such work was popular with the driver, who also got the traditional half-time hospitality of sandwiches, but it was also not unknown for them to be roped in to play if the team found itself a player short!

Pubs covered that season were the Rising Sun and Seven Stars (Blacknest), the Squirrel and Hernes Oak (Winkfield), the Gold Cup, Royal Hunt and Queen's Stag Hounds (all in North Ascot), the White Hart (Sunninghill), Bells of Ouseley (Old Windsor), Queen's Arms (Lovel Hill) and Duke of Connaught and Vansittart Arms (both in Windsor).

Various other clubs regularly hired buses to pursue their activities, the Ascot Police Sports & Social Club (visiting other forces for sports events and dances), Ascot & District Football Club, the Hatchet Lane FC, Virginia Water FC, Slough FC, Ascot Gasworks FC and Sunningdale FC, Cranbourne Cricket Club, Ascot Traders (to dances in Staines), the Police Fishing Club and Windsor Dorsal Fishing Club, various British Legion branches, the Ascot ex-Servicemen's Club, the Ascot Comrades Club, Fernhill Artisan Golf Club, the Sunninghill Old Scouts, Old Windsor Girl Guides, and the Post Office and Horlick's Social Clubs.

Most of these were to quite local venues, whereas other longer distance hires were organised by groups to the coastal resorts of Southsea, Brighton, Southend and Littlehampton. Other special events were the Aldershot Military Tattoo, pantomimes and theatres, whilst schools also featured with hires by Cranbourne, Sunninghill and the private St. George's at Ascot.

There were also some hire-ins by other operators, with *White's* (of Camberley), *Windsorian,* and *Smith's* (of Reading), the latter to assist with a larger transfer of Reading FC supporters to a local derby at Didcot.

There were also numerous hirings from within the Park community, some to the theatre or coast, others to Royal events in the town or to church services. Also operated during September 1938 was a special link from Ascot Station to the Royal Berkshire Golf Course, off Kings Ride, during an event there, and it is interesting to note that Harry Lintott had operated a service from Ascot and Bagshot Stations from 1934.

As can be seen there was a remarkable amount and variety of private hire operated by such a small outfit, given the commitments to the bus operations, and we will review how the nature of such work evolved over the years periodically throughout this account.

During 1938 there had been 73,730 miles run on the service routes, 5217 on contracts and 2438 on private hires, whilst 6486 gallons of petrol had been used.

1939 – Flight Of The Falcon

The early months of the year saw the little fleet suffer several mishaps on the road, the first of which involved 'Ace' (JB 9468) damaged during February, enough for it to have to be towed to Guildford for its repairs, which included a partial repaint. Details of the accident are not known, but apparently the insurers would not pay for the recovery work or entertain the hiring of another bus to cover its absence. Then on 5th March another bus was damaged by horses straying on the public highway, another incident the insurers disputed liability for!

In respect of the latter class of hazard, it should be appreciated that within the Great Park at the time there were far more large animals roaming more widely than in modern times, the deer covering a much wider area than the present deer park, whilst cattle were widespread, so the *White Bus* drivers all carried a stick for moving the beasts on, whilst a number of the gates had to be opened and closed with each passing, unless one of the Attendants was on hand. Indeed, although each gate had its official name, they were often known locally by the lodge-keeper's name, the latter tending to be there for many years, and indeed Cranbourne Gate was commonly called Butler's Gate!

This photo of the second Dennis 'Ace' (JB 9468) shows it outside the Seven Stars at the road junction by Blacknest, with the road to Cheapside to the left and that to the A30 and Virginia Water on the right. However, the real significance of this view is that in due course the new licensee was the mother of the future wife of Cecil Jeatt!

Comments have already been made regarding the general reliability of the Dennis buses, so when it was decided to invest in another vehicle that maker was again the choice, though it was also concluded that a higher capacity and coach seating would be required. In the meantime Dennis had introduced the 'Falcon' as a replacement for the 'Ace', though it had a more conventional wheelbase of 16ft 6ins, and as ordered had the same Dennis 4-cylinder petrol engine. The body was for 26 passengers with a front-entrance and built at Guildford, whilst it was specified that the sliding door needed to be opened by the driver when seated, as it would also be used on service routes at times.

The new vehicle was ordered in May for a July delivery, whilst that same month saw a March 1935 Austin 10hp car (CPE 957) purchased from Abbott's at Bagshot as the official vehicle of the Company.

Despite the looming war clouds the Aldershot Military Tattoo was prepared for June, though it would prove to be the last of that event, which was not continued after the war. *White Bus* ran 2 vehicles there on most evenings, although a Bagshot-based operator, *A.E. Butler* of By-pass Garage, took the opportunity of that year's license renewal to object to the pick up at Windlesham, but it was not altered. An exceptional hire also took place in June, when about 60 people from the Park were taken into Windsor for a Royal Procession, one bus doing three relay trips.

The swimming transfers for the County ran again from 19th June to 28th July 1939, which *White Bus* had secured by offering a 5% discount on its previous cost. The journeys ran each weekday except Friday, with the same schools and venues as noted before, though the Ascot Heath boys and girls were taken to different sessions, in this early example of fit-in work.

Figures for the first 6 months of 1939 reveal that 6832 passengers were carried on the Bagshot service, with daily loadings varying from 118 to 408, whilst 1589 of the passengers had return tickets. Also notable was that the daily fuel consumption varied from 14 to 24 gallons, indicating the use of a relief bus over parts of the route at times, with total fuel costs at £35 16s 8d and income at £122 4s 2d.

The Royal Show came to Windsor again in 1939, and was in fact the Centennial event, and of course would be the last for 6 years. It took place from 4th – 8th July in the Great Park, just off Sheet Street Road and accessed via Queen Anne's Gate.

There was quite a bit of correspondence over this event from a transport perspective with the Traffic Commissioner, as obviously *White Bus* wished to run extra journeys on its route from Windsor to Bagshot. However, it also soon became apparent that *London Transport* and *Thames Valley* had done a deal to have a jointly-operated shuttle service from the town to the site. Will stuck to his guns and corresponded directly with LT, as it was obvious they were in the driving seat over the arrangements, and claimed a share of the traffic for *White Bus.*

As a result the latter was paid £25 compensation for possible abstraction of traffic from Windsor, whilst the Board agreed to not object to his proposal to run up to four journeys per hour with a 20-seater from Windsor, whilst the link from Bagshot was also enhanced for those days. It also advised the Royal Agricultural Society to make up another board of 6ft x 3ft 6ins with 'Buses for Windsor & Bagshot' shown on the stand, whilst large paper notices on the buses proclaimed 'To And From The Royal Show'.

Here we see Vivie in Guildford High Street with the Austin car, shopping after perhaps visiting Dennis?

In the meantime a rumour had evidently started that *White Bus* was being sold, as during May Will had to ensure several local organisers that it was not the case. However, there was no smoke without fire, as an advert did appear in Commercial Motor on 26th July!

The ad only had a Box Number for responses, but stated 'Bus Services (Stage & Private) for sale, well established, Berkshire'. As far as can be ascertained there was only one response, when Cecil Hutfield, a Gosport-based bus, coach and garage proprietor wrote seeking fuller details, though no offer was to follow.

The 'Falcon' was taking shape over at Guildford when it was inspected on 14th July, though a revised date of 1st August was now quoted for delivery. Dennis was reminded that the Commer (JB 568) was offered in part-exchange, though it actually passed to the dealer Worthmores of London SW1 on the 1st September.

As soon as it arrived the Dennis 'Falcon' (BRX 865) was taken into the Great Park for these photographs. It had more brown paintwork than the 'Aces', with only a white waistrail, whilst the coach seating can be seen, as well as the sliding front doorway activated by a system of rods under the floor.

After a nervous few months the Second World War became a reality on 3rd September 1939, with changes to all aspects of daily life. For transport firms the following 6 years would be very difficult indeed, with rationing of fuel and rubber for tyres, whilst new or replacement vehicles would be virtually impossible to obtain. It was therefore rather fortunate for *White Bus* that they had just taken delivery of the 'Falcon', as well as having both 'Aces' overhauled.

Indeed, fuel rationing came into effect from Saturday 16th September, and from that same date the Traffic Commissioners called for a reduction in bus operations, generally resulting in the cutting of late journeys, some Sunday morning runs and longer distance private hire. Some more localised hires were permitted at first, though they were offset by additional work with troop movements and the transfer of evacuees starting from the outbreak of war.

The effects of the influx of mothers and children as evacuees to the area showed up in a comparison of the figures for the first full week in September, when 1687 passengers were carried against 1514 in 1938, such information being produced in support of fuel allocation claims required from all PSV operators by the Ministry of Supply. However, the fact that *White Bus* was not forced to withdraw further journeys does show how important its services were in view of the lack of other providers to those isolated communities.

The other immediate effects on bus operation was the 'black-out', whereby even the few lights existing along the routes were discontinued to deter use by bomb-aimers in locating targets, whilst most road signs were also removed for the duration to avoid them being used by the anticipated paratroopers.

There was rather less activity in the area in respect of 'shadow factories', with the established industries at the Slough Trading Estate, but the Park would play its part in the war effort. There was also a grass landing strip at Winkfield Plain, which became a divert field for RAF emergency landings, whilst various London-based Government Departments would find safer homes in some of the larger properties in the area.

1940/1 – Keep Calm, Carry On

When completing the fuel return for January 1940 Will asked for an allowance to be made for the taking of ladies from the Marie Louise Club to churches in the Ascot and Sunninghill are on Sunday mornings, but that was turned down.

On the other hand *White Bus* was approached in May to transport some 80 scouts and Leaders of the Fulham District Scouts to a Summer Camp at the Ascot Cottage Stables, near Crouch Lane, Winkfield, the organiser adding that they could walk from the public bus stop if required. The group was arriving at one of the Windsor Stations, but all their equipment would arrive by lorry. *White Bus* obliged by running a bus on a shuttle between those points until all were moved, charging 1 shilling per head.

Transfers of a more conventional nature continued for the County swimming contract during June and July, there now being an additional journey for the evacuee children at Cranbourne School, with buses running each day of the school week. It seems, however, that selling the business had not gone away, as a similar advert was re-inserted in early March, though there is no record of any responses with the war going on.

The Great Park was affected in a number of ways by the war, with the former Royal landing strip once used by the Prince of Wales when he resided at Fort Belvedere, improved to take heavier planes. Indeed, beside that on Smith's Lawn Vickers-Armstrong set up an aircraft works where bombers were constructed. The lake at Virginia Water was drained, with only the original Virginey Water stream left flowing, again due to its shape being a possible guide to navigators, even so over 200 bombs fell on the Park, but without any serious damage. During the Great War a large hut camp had been set up for the Canadian Forestry Corps and such activities were repeated for WW2.

The bus services continued throughout the war years, though spare parts became an issue and had to be requested via the M-o-S, whilst those who had cars before the war were obliged to lay them up, as fuel supplies dwindled. Sunninghill Park went to the RAF as Ascot Station, before being handed over to the United States Air Force, whilst nearby Silwood Park became Convalescent Depot No.118 from 1940. The local schools were swelled by evacuees, whilst as part of the Air Raid Precautions, the hospitals had been earmarked to receive casualties from London, with *Green Line* and *Thames Valley* coaches converted to ambulances retained at Windsor to meet trains.

Dennis 'Ace' JB 9468 at Bagshot Square fitted with headlamp masks and radiator muff during the war. It is not clear from this photo if the roof had been painted grey as was the case with many buses.

In May 1940 an enquiry came from the Office of the Crown Lands at Cromwell Road in London SW7 for a cost for a contract from Virginia Water Station to 'somewhere in the Great Park (not at liberty to divulge location) approximately 2.5 miles distance over perfectly good roads'. The letter sought a cost for a 20-seater with journey at 4.15pm and another an hour later to the station, without mention of morning journeys, whilst on Saturdays two runs at 12noon and 12.30pm were proposed. *White Bus* responded that it could do the journeys at £2 per day, but in the next letter were asked to provided a continuous service on one day only from either 10am to 5pm or just starting from 1pm instead. Although this did not lead to any immediate start it would do so later on.

Shortly before the war started it was found that the avenue of Elm trees flanking the Long Walk and planted in the reign of William III were diseased and required felling. During 1940 they were replaced by a mix of London Plane and Horse Chestnut.

Here we see 'Ace' JB 9468 taken from the mound of the Copper Horse looking towards the Castle along the Long Walk around 1940, the avenue of trees being replaced by saplings, so contrast this with page 90.

Horse-racing was also severely curtailed during the war, with some courses taken for other purposes and unnecessary travel discouraged. However, it was the plan to run The Derby at Newbury instead, so *Thames Valley* enquired through its Ascot Garage Inspector Frank Williams (who lived at Woodside) if *White Bus* could hire to them if required, though in the event the race was finally transferred to Newmarket for July.

A small amount of private hire was still evident in the Summer of 1940, mainly for the Cranbourne Cricket Club and only as far as Binfield, Broadmoor and White Waltham.

Much of the Great Park was closed off to the public from Tuesday 2nd July, though Eric Savill wrote to *White Bus* allowing them to continue operation, but only to 'residents and those with signed passes by me', also seemingly authorising the diversion of the Windsor – Bagshot journeys into the Park at Timber Hill Lodge as far as Chaplain's Lodge crossroads (about 1 mile along Dukes Ride), then back. As normal licensing provisions were suspended for the duration such a modification cannot be confirmed or otherwise, and may well have operated 'by request'?

King George III was a great exponent of agriculture, and during his reign Norfolk Farm and the adjoining Flemish Farm had been brought under cultivation, whilst much energy was put into road building and forming plantations, but in more recent times the Park had seen little farming activity other than fattening of cattle, though that would soon change due to the war.

The second 'Ace' (JB 9468) and the 'Falcon' are seen standing the North Street yard, with headlamp masks in place and the Hernes Oak pub to the rear left of them. Note also the white markings on the wings and below the radiator on the 'Ace' to aid visibility.

The contract run enquired about by the Crown Commissioners for Land started from Friday 24th January 1941, when it relocated its offices from Cromwell Road, West London to Fort Belvedere in the south-west corner of the Great Park. The bus left the junction of Chobham Road/London Road at 8.20am (close by Sunningdale Station), then to the station at Virginia Water for 8.35, with calls en route to Fort Belvedere at Christchurch Avenue/Virginia Water Road at 8.45 and The Wheatsheaf on the A30 at 8.50, before reaching the offices by 9am. At the end of the day it ran on weekdays at 4.10pm over that route, with a Saturday journey at 12.10pm, and £1 10s per day was paid for the service.

A little earlier a special one-off journey on 13th January had seen some staff from the Ministry of War Transport Insurance Office travel from Sunningdale Station to Buckhurst Park near Cheapdide.

With the war now definitely not ending in the shorter term, King George VI declared that the Royal Lands could not be an exception to the call for more home production of food, particularly as imported supplies from the Empire were now severely hampered by the German U-boat campaign. The deer in the Park were rounded up and restricted to certain areas, whilst 1000's of rabbits were cleared for agriculture, which mainly centred on Norfolk and Flemish Farms.

With the daily contract, all the Dennis vehicles were being kept busy, but on the evening of Thursday 17th April a local private hire was undertaken for the Humglas Social Club based at Winkfield Manor, the 'Falcon' conveying them to Windsor Theatre Royal.

However, as we have already noted, Will Jeatt was not in the best of health, and on 6th May 1941 he passed away at the King Edward VII Hospital in St. Leonards Road, Windsor. His wife Beatrice was still out of the area, having taken the post of nurse at St. Edward's School in Dorset, and anyway had not been a partner in the bus operations, a fact confirmed to Cecil and Vivien by their Solicitors. That left them and driver George Jacobs to 'keep calm, carry on' in keeping with the slogan of the day, and George is recalled for his rather unique driving style, as he seemed to be constantly adjusting the steering wheel, even on a straight road!

Despite the generally difficult situation for transport, the County Education Committee did continue with the swimming lessons during June and July 1941 as before. However, a further request for fuel allowance for taking the ladies of the Marie Louise Club to Sunday-morning church services was not upheld in June 1941, at which time the buses were running 2786 miles per week and consuming 280 gallons of fuel.

In response to issues raised by shortages of fuel and reduced lighting, the Windsor Police decided to alter the stopping and terminal points for buses in the town centre, so *White Bus* was moved into the GWR Station from 17th September 1941, joining the *Thames Valley* buses, most of which moved there back in 1931.

Unfortunately 'Ace' JB 9468 was involved in a smash on the A30 between Sunningdale and Bagshot at 8.45 am on 22nd December 1941, colliding with the rear of a stationary lorry which showed no rear lights.

1942-6 – Make-Do-And-Mend

1942 saw many supply situations deepening further, as available raw materials were swallowed up by the war effort, whilst there was virtually no prospect for new buses. The Ministry of Supply had authorised the completion of some in the course of construction or made up of parts in stock, and also some intended for export had been diverted to home operators, but only limited numbers of double-deck Guys and Daimlers, along with single-deck Bedford OWB's were built from new and allocated by the Ministry.

During February 1942 the Ministry invited operators to request spare parts that would result in immobile buses returning to use, whilst under wartime conditions it became an offence to buy or sell any bus or coach without authority. Another communication of May 1942 also invited all operators to consider the running of buses on gas, though no *White Bus* did so.

The Marie Louise Club had the King and Queen as its patrons, and in June 1942 its Secretary wrote to say that the Ministry of War Transport had now approved a fuel allowance increase to cover the Sunday church journeys, which could then resume.

With the sorting out of Will Jeatt's financial affairs in November 1942, it emerged that he had not kept up the 'stamps' for National Insurance for himself and George Jacobs, the latter then nearing retirement age. Although he was generally quite efficient looking after the office paperwork, he had not provided enough stamps to qualify either of them for a pension payment. To complicate matters, and because of his previous bankruptcy, he was regarded as an employee of his son and daughter, so they got a hard time for not ensuring compliance, and indeed were lucky not to be fined! In the end they were allowed to 'stamp up' George Jacobs so he would qualify, but in the case of Will no pension allowance would be made, which also meant that Beatrice had no entitlement either.

Indeed, Will's financial troubles did not even end with his death, as in February 1943 Cecil got a letter from the Official Receiver pointing out that furniture and a gold watch inherited from his father was considered as due as payment off the outstanding bankruptcy claims. In the end Cecil was allowed to send some cash instead and keep the items of sentimental value.

By December 1942 the fuel situation had bitten harder, so *White Bus* requested the cessation of a later journey, as well as all Sunday operations, which was approved from 12th January 1943.

Vivien was preparing to marry Eric ('Dick') Mauler in the Summer of 1943, but the Spring would prove a trying time for the brother-and-sister outfit. As it was Dick was serving in the Forces and we shall hear more of him in due course, whilst Cecil had been in the Home Guard since June 1940.

However, the Spring of 1943 started off rather badly for *White Bus Services* on the vehicle front. On 2nd April the second 'Ace' (JB 9468) was involved in a bad smash out at White Waltham, as a result of which it had been recovered by H. Markham Ltd., the Reading coachbuilder based in the old *Thames Valley* garage at 113 Caversham Road. The work involved stripping the nearside, rebuilding the framing, replacing the front door and panels, straightening the side louvres and windscreen frame, and re-glazing the windscreen and side windows, and such was the task under wartime shortages that it was not back on the road until 19th September!

Also that month the first 'Ace' (JB 4838) received a check by the PSV Examiner, as a result of which a 'stop notice' was served, which meant it could not be used until the items listed were rectified. The report noted the 'body to be in dangerous condition, with rotten pillars and extensive body movement, and door hangers loose'. Clearly there was little scope for putting all this right short of re-bodying, especially as the other 'Ace' was away with no return date in sight!

Not for the first time in its history *White Bus* put out urgent enquiries, and indeed it was lucky to obtain a 1931 Dennis 'Dart' (KX 7454), with permission of the M-o-S from *Walter Oborne* of Aylesbury for £750 on 26th April 1943. It had actually been re-bodied by Duple in 1939 with a 20-seater coach body, so was a sound vehicle, the model having a 13ft 2.75ins wheel-base and a 6-cylinder 26.9hp Dennis petrol engine.

The Dennis 'Dart' (KX 7454) seen at Bagshot Square.

In early 1944 Cecil Jeatt married Violet Mahoney, who had been born in Bermondsey, South London in 1916. Her parents had a pub near Hay's Wharf and moved to the Seven Stars at Blacknest to avoid the blitz, Cecil meeting her when she was employed by

Colonel Horlicks at Tittenhurst Park as an unofficial land-army girl looking after cows. They had daughter Christine later that year, who duly passed her PSV as a Driver and did some work for *Winkfield Coaches* in the 1980's, with her sister Patricia following in 1946.

The other significant event of 1944 was the Education Act, which included legislation to increase the school leaving age to 15, but of more direct relevance also set out the provisions for Local Education Authorities in their responsibilities for home-to-school transport, and indeed much of that still stands today.

Before the war most children received their schooling near to home, but after then the trend started towards closing smaller local schools down in favour of larger ones, followed by the post-war 'baby boom', which placed quite a strain on the established larger schools as well, both factors leading to a huge increase in the need for home-to-school transport especially for the secondary years. Indeed, in Berkshire the new schools tended to be built in 'neutral' areas rather in a town, the catchment areas for each radiating out some way. The planning of Further Education also underwent changes, so many pupils would have to travel from age 11 through to 16 to school and then onto college.

It is worth just taking a look at the main criteria laid down by the Act, in as such as they affected the way transport developed. For pupils in Primary Schools a walking distance in excess of 2 miles was required for eligibility for free transport, which might actually mean a bus pass on a public service route, the modern paranoia not then in place regarding children going on their own. For Secondary pupils the distance had to be in excess of 3 miles, though in both cases there could be exceptions where the walking route was deemed unsafe, though again the latter term was not based on the parental interpretation as, in better keeping with the times, parents did then commonly take children to and from school in association with neighbours etc.

To give a couple of examples from my own dealings with home-to-school transport, the first concerning a brother and sister living in Wokingham District at the junction of Bearwood Road and Barkham Road, each allocated to single-sex schools, with the boy at The Forest and his sister at The Holt. Neither school was over 3 miles from their home, but the boy's walking route was deemed unsafe because of a bad history of accidents and no continuous footpath, so he was given a free pass on a contract route. However, the girl had a safe walking route of some 2 miles, so was not eligible to a free pass. As this was a widespread issue along that road a former contracted route had been allowed to go over to a registered school service, with the operator selling passes to those not entitled.

In the second example, actually in the Great Park, the officers went out and measured various distances for

eligibility, but after appeal some passes had to be issued as free because, although the traffic was severely restricted, there was possible danger from rutting deer!

Although the decisions of the Authorities may not always be welcomed by individual parents, these rules did make it necessary for such needs to be met by the Education Departments, though whether by utilising public bus (or train) services or contracting routes did depend on the actual journey. Also to be borne in mind is that many larger bus operators actually started to reduce service networks from the mid-1950's, with a similar reduction on the local rail lines in some areas, which all led to more contracted school buses.

I have spent some time on this topic because of the difference such provisions would make to the survival of *White Bus*, and indeed such work was the salvation of many a small operator in those post-war years, as it offered a 5-day-a-week job for some 38 weeks a year, but left weekends and the major coaching seasons of the Summer and Bank Holidays with vehicles free.

The first Dennis 'Ace' (JB 4838) would see no more use after being served with the 'stop' notice, but did survive for some years as a source of spare parts at the North Street premises.

The only other notable event affecting *White Bus* in 1944 was due to the increased use of the Great Park for agriculture, which meant restricting the areas where cattle could access. From 19th May all the gates other than those on the main road (which had cattle-grids) were to be kept closed, which meant that each time the bus passed them they needed to be opened and then closed again by hand.

Although the D-Day Landings had foreshadowed the return of peace, the shortages of all materials would continue for many years, and even the final food rationing would not cease until 4th July 1954!

By early 1945 the Dennis 'Falcon' (BRX 865) was in need of a thorough overhaul, having worked hard since it had arrived back in August 1939, so it was sent to Guildford in mid-March. It was fitted with a re-conditioned replacement engine, whilst the other work amounted to four long sheets of parts used to bring the invoice up to a total of £173 12s 5d.

When the 'Falcon' returned from overhaul it was re-painted in maroon and white, and was seen at Bagshot Square by John Gillham in April 1945 in that scheme.

The war in Europe ended on 8th May 1945, celebrated as VE Day, but in the East the surrender of Japan did not come until 14th August 1945. However, neither of these events brought anything like a return to normal life, whilst it would take some time for all the serving personnel of the Armed Forces to return home.

The black-out restrictions had been dropped earlier, but after the D-Day Landings what oil could be imported largely went on the final drive to victory, so the fuel situation saw no immediate improvement. Indeed, the restriction on mileages for private hire work was not removed until April 1946.

In the meantime in August 1945 the opportunity was taken to purchase the North Street site from Mr. Willett, who lived in Wales and had inherited the land with the death of Mr. Willis just after signing in 1932.

Also with the war now ended an appeal was made for the release of Corporal Dick Mauler (out-of-turn), as he was needed for essential transport work. He had provided motor-cycle escort for the large Queen Mary Transporters for the Royal Air Force when taking planes by road, and his maintenance training would certainly come in useful for keeping the fleet running.

As a consequence of the war ending, the daily contract for the Crown Commissioners ceased from Saturday 12th January 1946, and the letter confirming that made mention of the efficient service provided for its staff for the past 5 years.

During June 1946 a letter was received from the Windsor RDC seeking a better service for residents of the Cheapside area, and a similar one had also been sent to *Thames Valley*. The latter was actually having problems covering its own services at that time due to a very war-worn fleet, and it would not be until the following Spring that full pre-war levels of services were once again attained. *White Bus* did of course re-instate the later journeys and Sunday operations as soon as fuel permitted, though no date has come to light for when that was actually achieved.

The family consolidated its position with domestic accommodation from August 1946 when another property in Crouch Lane was purchased. The Jeatt family had of course occupied No.1 for some years, but now Vivie, Dick and Christine Mauler stayed there, whilst Cecil and his wife, bought the former bakery close by and re-named it as 'Puck's Blarney'.

In a reminder that Drivers also needed to understand their charges, Mrs. Mauler effects some minor adjustments to the engine of Dennis 'Ace' JB 9468, still fitted with headlamp masks.

1947-9 - First Bedford Arrives

Little occurred at *White Bus Services* during 1947, but the arrival of Douglas Edouard Rule Jeatt into the world would of course prove to be significant to the story much later on.

The swimming transport was further expanded for that June and July by the addition of pupils from Binfield Church of England, Warfield C.of E., Winkfield St. Mary's C.of E. and Priestwood School in Bracknell all being taken to the pool at Bracknell, a total of 158 children in all and over 5 days each week.

There were no changes to the fleet until April 1948, when the first of many vehicles of Bedford chassis was purchased from Arlington Motors of Vauxhall Bridge Road in London SW1, being a WTB-type built in July 1939 before the war prevented many more passenger vehicles from being completed.

The Bedford WTB (BUN 677) carried a Duple 25-seater body of the 'Hendonian' style, a type specifically intended for the small bus operator wanting to use the same vehicle for private hire work as well as stage carriage. The later OB-type radiator grille was carried on so late an example, as seen here at The Square in Bagshot in this photo by D.A. Jones. It had been new to Phillips of Rhostyllen, but came via several other operators to Winkfield.

BUN 677 saw use widely on the services and private hire, and by November 1948 had its seating altered to take 26 passengers, as well as 8 standing on bus work.

At the start of 1948 some regular hires had started to *A. E. Butler (Cove Saloon Coaches),* now of By-pass Garage in Bagshot and Victoria Road in Cove. Also worthy of note is that a 20-seater day hire to Southsea was charged at £8 15s and for a 26-seater was £11 7s 6d at that time.

From 29th July to 14th August 1948 Britain hosted the first Olympic Games since that staged in Berlin in 1936. The next in 1940 had been scheduled for Tokyo and that for 1944 would have been London, but of course neither occurred, whilst neither Germany nor Japan got an invitation in 1948! Also known as the 'Austerity Olympics', no new buildings were built for the event, and athletes were housed in existing places, for the shortage of materials was still keenly felt. In respect of the local area, the rowing events were held at Henley-on-Thames, whilst the two men's cycling road races took place in the Great Park, bringing more visitors but also disrupting some bus journeys.

Another significant factor of 1948 was the start of the Model Village, not a replica English village such as that at Bekonscot, but one of groups of cottages around Richardson's Lawn and nearby The Prince Consort Workshops. These were model in the sense of being good quality homes for the Park workers and their families, and the initial 32 homes were matched by a Post Office Stores. They were added to in 1954 with 18 more and another 6 in 1966, though the effect was quite time-less, with the village green setting and nearby Shop Pond and Isle-of-Wight Pond. It was the largest single development within the Park and would have an important influence on the bus service, with generations of its residents utilising it to travel into town for shopping or leisure, school and college.

Private hire work for 1948 had increased markedly, with 102 jobs in all, varying from local dart teams to full-day private hires to the coast, with The Fox pub at Winkfield Row the most numerous at 12 hires.

Some mention has already been made of the regular hires from within the Park community, and a Branch of the Women's Institute had existed since 1932, the same founding year of the Football Club, though pre-dated by the Cricket Club begun in 1861. Bowls and Tennis Clubs had followed in 1954, then a Golf Club in 1978, all of which would naturally look to the local firm to provide transport to events or away fixtures.

In respect of the social facilities in the Park, the former Vickers-Armstrong work's canteen near Egham Wick was dismantled and its wooden frame re-erected as the basis of the York Hall, which opened near The Village in 1951 as the new social centre in place of old converted stables previously used, a good use of materials in such times of scarcity, and another example of how the workers and families were not forgotten by those in charge.

The *White Bus* finances were, however, far from good, no doubt a factor in Mrs.V. Jeatt's decision to take a part-time teaching post at Ascot Heath School from March 1948. Indeed, she had previously been a pupil at Notre Dame Convent in Southwark, South London, before training as a teacher at La Saint Union in Southampton, duly becoming a supply teacher with London County Council from 1936. In due course she would move to St. Ethelbert's RC School in Slough in January 1953, and was at St. Edward's RC School in Windsor by Autumn 1961, amongst other local Catholic schools. Indeed, her faith and consequent employment, along with the schools attended by her own children, would make her well known throughout the Roman Catholic communities based in the Ascot/Windsor/Slough areas, which in turn would have a significant impact on future private hires and school transport undertaken by *White Bus Services.*

On the other hand, there is little doubt that without her salary as a teacher the family could not have survived on the proceeds of *White Bus* operations, a fact Doug Jeatt acknowledges when he says 'he doesn't know how they did it, raising four children through at times privately funded education', though he adds, that 'all his parents did was work and raise the family'. She even paid for several vehicles out of her income when urgent replacements outstripped the Company funds, and those daily users of the buses had no idea how close it came to losing their services at such times!

Although the Bedford WTB had been added in April 1948, it was not until July 1949 that the Dennis 'Dart' (KX 7454) was disposed of after 6 years in service.

For the Autumn Term of 1949 *White Bus* was asked to tender for a daily school transport for Princess Margaret Rose School in Vansittart Road, Windsor, and despite that coming to nothing, such work would feature more in due course.

White Bus Services was a member of the P.V.O.A, which represented the interests of its members with legal advice and matters relating to legislation.

PASSENGER VEHICLE OPERATORS ASSN. LTD.
ROADWAY HOUSE,
146 NEW BOND STREET, LONDON, W.1.
———
This is to certify that
White Bus Services
(Messrs. C. E. Jeatt & V. M. Mauler)
North St, Winkfield, Windsor.
is a member of the Association for the year
ended 1st October 1946 BERKS
Area Secretary.

At the close of 1949 the *White Bus* fleet consisted of the 1936 20-seater Dennis 'Ace' (JB 9468), the 1939

Dennis 'Falcon' (BRX 865) and the 1939 Bedford WTB (BUN 677), both of which seated 26 passengers, the 'Ace' being quite scarce nationally by then.

1950-2 – Festival of Britain

The dawning of a new decade brought little respite from austerity, so the Government commissioned an event to give the Nation something to celebrate on the centenary of the Great Exhibition of 1851, as the Festival of Britain set for 1951, of which we shall more shortly.

On the *White Bus Services* an extra journey was added from April 1950 to give a wider spread of timings on from Windsor (Western Region Station) – The Crispin Hotel of the existing 11am and 3pm, with a 7pm run, which had reached Windsor as a short-working from Sunninghill at 6.15pm. That bus ran through the Park to The Crispin, and then came off service to return to North Street.

A Commer 'Commando' was acquired via Arlington Motors (EUX 524), which had been new to *Hoggins, Wrockwardine* in December 1948, and it carried a 30-seater, front-entrance Harrington coach body.

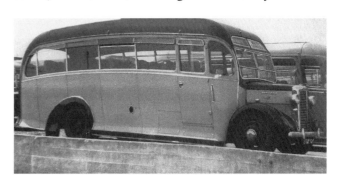

The Commer (EUX 524) had a good-quality body by Harrington, and the model was at that time one of the competitors to the 28hp Bedford OB. It was photographed at the Southsea Coach Park in this shot courtesy of Phil Moth.

The arrival of the Commer heralded the demise of the second Dennis 'Ace' (JB 9468), which had been new in July 1936 and was withdrawn in August 1950. The 'Ace' was something of a rarity by then, but this vehicle had three subsequent owners. However, the first of the 'Aces' (JB 4838) was not so lucky, having already been stripped of parts since its withdrawal back in 1943, it went for scrap instead.

As it happened, the Dennis 'Falcon' was involved in an accident around the same time, so *White Bus* took an insurance payment of £500 in lieu of repairs on 17th August, giving another headache of finding a replacement vehicle at short notice. It also meant that for the first time since September 1933 the Dennis make was not represented in the fleet.

One last shot of the second Dennis 'Ace' (JB 9468) in the 1937 Coronation year at Forest Lodge. It had served White Bus well over 14 years, especially when considering that many larger operators considered a 7-year working life acceptable for service buses.

In order to replace the 'Falcon', which was sold for scrap, a Bedford OWB (CMO 495) new in October 1942 with a 'utility' Mulliner 32-seater front-entrance bus body was acquired from *Windsorian Coaches,* a few miles away. The official transfer date is recorded as November 1950, but it is understood to have been on loan from August, both firms being on good terms.

The utility lines of bodywork on the Bedford OWB (CMO 495) are evident in this D.A. Jones shot at Windsor (WR) Station. It still has the slatted wooden seats, but headlamps larger than originally fitted. The livery is pale grey with bonnet top, waistrail and mudguards in red, which was a version of Windsorian's scheme. The belt-and-buckle device on the body sides is only known from this bus, with 'White Bus Services Phone Winkfield Row 4273'.

October 1950 saw several representations for buses on various fronts. One came from Cheapside residents in respect of a direct link to Ascot High Street, whilst the other was from the Lightwater Conservative Group asking the Company to consider stepping in after the withdrawal by *Aldershot & District* of its Route 58 (Weybridge – Camberley), but nothing came of either matter at that time.

A further suggestion came from Sunninghill Parish Council in January 1951, asking if buses could be diverted at the Berystede Hotel to run through South Ascot – Ascot Station (SR) – Ascot (Horse & Groom) – Royal Hotel – Swinley Road and thence to Bagshot in similar fashion to the abortive 1938 proposal, but again no changes were made at that time, though some elements do of course form the current route.

As an indication of how private hire had developed, during the first 3 months of 1951 some 2399 miles were run, carrying 1664 passengers and bringing in £190 income. In January they consisted of 31 jobs on 30 days, sometimes with school-related work during the day, followed by social activities in the evening. A number of entertainment venues featured, mainly for pantomimes, such as the Chiswick Empire and the Empire Hall in Wembley, whilst the football matches of Reading FC, Maidenhead FC and Ascot FC were catered for. Various groups from the Great Park used the Company for sporting fixtures, to visit other W.I. branches or to regular whist-drives and dances, whilst most of the local pubs booked for dart matches.

The bookings for February 1951 added more variety, with football at Camberley now included, it being the local team for the Bagshot area, whilst out of 27 hires over 23 days one even featured a school attending the funeral of a member of staff.

By June and July the Summer outings to places much further away were evident, with journeys to Bognor Regis, Bournemouth, Littlehampton and Margate, as well as Richmond for ice-skating, Oxford for the then very popular Speedway races, as well as covering the race card at Epsom and Goodwood, the Ascot area pubs in particular being keen followers of the horses.

These jobs and the service journeys were covered by family members Cecil Jeatt, Dick and Vivien Mauler, aided by drivers known only as Don and Sid along with the retired George Jacobs on some excursions.

The family of Cecil and Violet continued to grow, with Katherine (1949), Matthew (1955) and Gerard (1957), all of whom we will hear more of later on.

However, the main event of 1951 was the Festival of Britain, which opened on 3rd May, which was also the day the author came into the world! The exhibition was mainly centred on the South Bank, near Waterloo Station, on a large bomb-site area, where pavilions housed themed exhibitions celebrating Britain's past, present and future, with an emphasis on developments in technology, only the Festival Hall now remaining.

The fun side was also evident, with the Battersea Fun Fair just a short way upstream, whilst numerous other local events took place across the land, and a floating exhibition toured ports during that Summer.

At some point in 1951, as yet undiscovered due to motor tax records being incomplete, a further Bedford vehicle was acquired in the shape of another OWB (ATL 835) but carrying a Duple 30-seater front-entrance body to 'relaxed' standards, including upholstered seats. It had been new in January 1946 to *Delaine Coaches* based in Bourne, Lincolnshire, though running in a pale cream livery rather than the ornate blue and cream scheme they are known for.

Seen by Bob Mack at the old Post Office in Windsor, Bedford OWB (ATL 835) was actually the first Bedford PSV in the country to be fitted with a Perkins P6 diesel engine, and also the first of that mode of propulsion at White Bus. The badge on the radiator was of the engine maker based at Peterborough.

Mrs. Jeatt (as she was usually known, not liking her forenames) had an Austin car before marrying, but in October 1951 got a Fiat 4-seater (MP 271) originally registered in May 1927. She would duly replace that with a 1934 Austin 9.9hp saloon (BBH 404) in 1952, followed by 1936 Austin 11.9hp (DMY 651) in 1956, then a 1951 Vanguard estate (KKV 567) from 1960, and a 1955 Humber 'Hawk' (606 DMH) in 1964.

By early 1952 King George VI was very ill, the war years having certainly taken a toll on his health, and on 6th February 1952 he passed away aged just 56. As it was that was not a good time for the health of Cecil Jeatt either, as he suffered from Sarcoidosis, which affected his lungs. He attended Brompton Hospital at Fulham, West London that year, with treatment over the next 4 years at least, which of course meant his enforced absence from the business at times. He also had a large number of x-rays taken, and that probably contributed to the Leukemia which shortened his life.

The fleet nearly had some interesting additions from 1952, when an advert in Commercial Motor was responded to for a pair of 31-seater coaches on Vulcan PPF/1-type chassis, a quite rare marque by that point in time. These vehicles (KTF 672/4) were new in 1949 and offered by the *Grange Motor & Cycle Co. (Lake District Motor Tours)* of Grange-over-Sands, who had sold out to *Ribble M.S.* and were selling these quite recent vehicles.

July 1952 saw some further modifications to the Windsor – Bagshot service, which added some short-workings between The Crispin and Sunninghill High Street, which were aimed at meeting the desires of the local Parish Councils, as most day-to-day purchases could be obtained in the shops there, the additional journeys operating daily except Sundays.

However, on the other hand, the last journey on Saturdays and Sundays from Windsor was now cut to run as far as Sunninghill only, but worked back to The Crispin before a dead trip to the North Street base.

In respect of that base, in August 1952 permission was sought to re-site the petrol pumps by the road-side in order to offer a public filling facility, but due to there being insufficient pulling-in space the Easthampstead RDC turned the proposal down.

Cecil Jeatt as seen in his mid-1950's passport photo.

1953-5 – Parting Of The Ways

Queen Elizabeth II succeeded her father to the throne, and the Coronation was set for Westminster Abbey on 2nd June 1953. Huge crowds flocked to see the Royal procession, but apart from that day of celebrations, there were the associated illuminations, all of which generated much excursion traffic, with a number of local groups booking trips, including the members of the Royal Household at Windsor Castle.

Despite that it was generally a quiet year for any changes for the *White Bus* fleet or services, the fleet remaining as the quartet of Bedford WTB-type (BUN 677), the Bedford OWB-types (ATL 835 and CMO 495) and the Commer 'Commando' (EUX 524), as well as the Austin car (CPE 957).

In the meantime the desire to operate a petrol-filling station had not gone away, with a representative from Power Petroleum going to see Berkshire CC (as the Highway Authority) to approve the plan, but they said they could not recommend approval because there was Smart's Petrol Station almost next door and another filling facility across the road. The Smart's we will hear more of later, being the same family of Billy Smart's Circus fame, and in due course a Safari Park.

Following that disappointment it was decided to re-arrange the in-house facilities with an order for a new Avery-Hardoll Model FP hand-operated pump, plus a 600-gallon above-ground storage tank, all ordered on hire-purchase from Power Petroleum of Charlton, London SE7 in March 1954.

Private hires for 1954 show a continuation of themes noted earlier, with regular dances, whist-drives and darts the most prevalent evening jobs, Richmond had a very popular ice-rink, whilst being so close to Ascot Racecourse there was inevitably a strong following locally for the gee-gees, with many local pub outings to Epsom and Goodwood in particular. One exceptional hire during 1954 involved 2x29-seaters and a 30-seater for a local golf tournament.

In connection with operations and costs, in April 1954 wages were being paid to Cecil Jeatt, Vivien Mauler, Wally Freeman and Franie Roche, the latter also being a lorry driver with Harry A. Coff on gravel and sand haulage based at Cheapside, one of several from there to do part-time or go full-time onto *White Bus* over the years. Each of the above got £7 5s 9d for a week's work, though Cecil only took £5 for himself, whilst the average earnings of both bus services was £80 per week at June 1954. During the Summer there are additional payments to 'driver', amended to Dick Mauler from September.

Further insight into the daily activities of *White Bus* comes from Michael Clancy, who as a young boy was living at Hilltop Close, Cheapside in the 1950's. He travelled to his nursery near The Berystede or to Windsor with his mother, recalling that Cecil and Vivien were always happy and smiling, which pleased the passengers, whilst they knew them so well they would set them down at their cottage doors in the Park as required, all much appreciated with the groceries to carry home, and an exciting change for the young boy to see other parts of the Park.

On the first afternoon run from Sunninghill, there would be a change of drivers at The Crispin, which often meant Cecil handing over from covering the early turn to his sister Vivien, after which he went off for his lunch. Michael also recalls seeing the bus parked in Windsor WR Station during the lay-over and just behind the *Thames Valley* stands, with the driver round the corner supping tea in the little café.

In March 1955 a Bedford OB with Mulliner 28-seater front-entrance body and new in 1947 (DRX 296) was acquired from *Frowen & Hill (Borough Bus Service)* which had sold its Windsor - Clewer Green –Windsor circular local service to *Thames Valley* on 1st December 1954.

The Bedford OB (DRX 296) from Borough Bus is seen in the Windsor WR Station, with its distinctive advert panels situated opposite the bus stands and seen by 1000's of people daily, including one for Stratford-on-Avon. The bus still wears the two-tone red livery of its former owner in this Bob Mack photo.

Further service changes took effect from 1st July 1955, one adding a mid-day short-working from Woodside to Sunninghill and back. A more fundamental change saw 4 buses per day diverted through Windlesham village as opposed to stopping at The Windmill on the A30 main road, presumably to increase patronage.

We have seen how Cecil and Vivie had come into the business practically from childhood, then since their father's death had jointly run *White Bus Services,* but by the Autumn of 1955 a decision was made to split the operations from October.

Under the new arrangements Vivie and her husband Dick Mauler would operate as *Winkfield Coaches* to take on private hire and contract work, whilst Cecil would retain the bus services, the Ascot Races extra journeys and the Berkshire CC swimming contract. An unwritten agreement apparently precluded Cecil from actively seeking private hires, but inevitably some locals stuck with him, whilst the contact with the Park residents 'rubbed off' to provide continuing hires from that community.

We will return to the subject of *Winkfield Coaches* in a later chapter to follow progress seperately, but they now took over the Commer 'Commando' (EUX 524) and the Bedford WTB (BUN 677), both of those being coach-seated, whilst Cecil retained a Bedford OWB (ATL 835) and an OB (DRX 296), with the other OWB (CMO 495) having been sold in September 1955 following the arrival of its replacement from *Borough Bus.*

Both operators still used the same site, but the coaches now occupied the covered garage, whilst the buses stood on the open ground to the right. The existing phone number (Winkfield Row 4273) was used by *Winkfield Coaches,* with Winkfield Row 112 allocated to *White Bus Services,* whilst the offices for each were respectively now at 1 Crouch Lane and 'Puck's Blarney'. In reality each often covered work for each other, with maintenance involving Cecil and Dick.

Given the quite precarious state of the *White Bus* finances, its continuation from that juncture was by no means assured, and local passengers never knew how close they came to losing the services altogether!

During September 1955 an application had been put forward to increase the frequency of the Windsor – Ascot Racecourse 'express' service to be on-demand, but *Windsorian Coaches* wished to object. However, Don Try, Manager of the latter wrote to Cecil to say that if they moderated the expansion to 4x29-seaters in each direction daily he would not object, adding 'will pick you up for the Hearing at 9.45', which shows how local operators often got along well.

By the end of 1955 a total of 136,965 passengers had been carried, over a total of 107,617 miles, and of course the hand-clipped pre-printed Bell Punch tickets were still the order of the day at *White Bus.*

1956-8 – Fuel Rationed Again

By the Summer of 1956 special journeys were put on for the polo season between May and September, with a bus running at appropriate times from Windsor (WR Station) to Smith's Lawn each Sunday.

During that Summer several changes took place to the *White Bus* fleet, resulting in the withdrawal of the OB from *Borough Bus* (DRX 296) after a relatively short stay in September though it continued in public use, albeit for the Cypriot operator *Kyriakola & Sons* of Limassol and re-registered as TAA 754!

Incoming Bedford OB (MVX 508) was caught by John Gillham parked in Crouch Lane in June 1961.

Another Bedford OB, but new in 1948 and with a 29-seater coach-seated front-entrance Duple 'Vista' body (MVX 508) arrived in July, having originally been with *Essex County Coaches,* but coming via another.

That was joined in August by something really rather different, and indeed the only normal-control Leyland ever owned by *White Bus.* It was a 1950 'Comet' CPO1-type, with a 5.8 litre oil engine and 5-speed gearbox, which gave a lively performance, and featured vacuum-assisted hydraulic braking. Although the type was not built in great numbers for the home market as a PSV, its 17ft 6ins wheelbase allowed the Windover front-entrance coach body to seat 33. This example (KXU 673) was one of a batch of 6 new to *Birch Bros.* of Kentish Town, North London, and it retained their cream with green livery.

A photo of another of the Birch Bros. 'Comets' (KXU 672), showing the American influence on frontal styling and 'crocodile' bonnet top for engine access.

With fuel rationing now thought to be a thing of the past, the unfolding Suez Crises in Egypt saw Israeli troops entering the canal zone, apparently with the blessing of Britain and France, which threatened to involve the post-war Cold War super-powers of the U.S. and U.S.S.R. It also highlighted the short-falls of the West's dependence on Middle East oil supplies, a factor which still shapes much foreign policy to the present day. Supplies dried up through the canal, so fuel rationing was again introduced for far beyond the actual duration of the fighting between 29th October and 7th November 1956, with the Israelis not leaving the area until March 1957 and the canal back in use from April. The bus services ran as usual, but some private hire had to be turned down as a result.

However, an exceptional local hire did see 50 people picked up from the Park on 24th December 1956 and taken to the splendid surroundings of St. George's Chapel in Windsor Castle for the annual carol concert. It is also worth noting that between June and October *Winkfield Coaches* covered some bus duties 14 times.

A further acquisition from *Borough Bus* followed in October 1956, though it was a utility-bodied Bedford OWB (CMO 773) new in 1943, but by then had been fitted with 26 upholstered seats.

Above - Bedford OWB (CMO 773) is shown in early post-war when still with Borough Bus, in a photo by Joe Higham at the Park Street terminus then used for certain services. Below - We see it later with White Bus, showing that it was repainted white and green, when caught in Sunninghill Hill Street. The Hovis sign is still over the Anne Marie Bakery there.

On the services numbers were up by almost 10,000 for 1956 over the previous year, with mileage nearly 1000 miles less, both factors probably reflecting the effects of the fuel crisis, as motorists took to the bus instead. Analysis for November shows that 3027 used the Windsor – Crispin route at £88 12s income, whilst the Bagshot service carried 9443 passengers for £288 12s. As a result of the arrival of the above Bedford, the Perkins-engined example (ATL 835) was let go in March 1957, though it remained in use locally with *A. Cole (Blus Bus Service)* of Slough.

A further expansion of the Berkshire CC swimming contract took place from June 1957, with St. Francis R.C. School, South Ascot also added. Also that Summer the Windsor Great Park Bowl's Club started to hire on a regular basis for its away fixtures, another example of the loyalty of that community to Cecil and *White Bus*. On the flip-side of the coin, there were again 14 days when *Winkfield Coaches* covered a bus working from January to July 1957, perhaps reflecting one of Cecil's absences through ill health?

Indeed, there was much contact between the two firms, as sharing the same site meant that cover could quickly be arranged, whilst it is noted that driver Franie Roche, can be found doing part-time turns for both, sometimes using a vehicle borrowed from the other or a job subbed out! *White Bus* drivers of that time also included Ray Cleeton, another part-timer.

However, one quite exceptional hiring offered to Cecil did not feature his sister and brother-in-law, though it entailed assistance from another source. The task was on Thursday 18th July 1957 and involved taking 3 coach-loads of pupils and teachers from Sunninghill School on a 200-mile circular adventure, which took in the sights of Winchester – Romsey – Bournemouth – Poole – Wimborne Minster – Blandford Forum – Salisbury – Stonehenge – Andover – Whitchurch – and back into Sunninghill through Bagshot. Coaches 1 and 2 were to leave at 30-minute intervals on the route as described above, whilst the third was to take the circle in the opposite direction, with Bournemouth to be regarded as the half-way point for a picnic on the beach, whilst other stops were to be at the discretion of the staff. The whole job cost £45, with two coaches being hired from *White's of Camberley.*

We have already seen how the old GWR station at Windsor had become the Western Region under British Railways, and then from August 1957 it changed once more to Windsor Central Station, the timetables reflecting such facts over the years.

The tickets used were still the pre-printed hand-clipped type, and the order placed with Bell Punch in September 1957 reveals a purchase of 80,000 tickets, split into bundles of 3d pink, 7d lilac, 8d yellow, 9d buff, 10d green, 1s purple, 1s 2d apple green, 1s 3d primrose, 1s 5d brown, 1s 8d blue, 2s magenta, 2s 8d white, 2s 11d green, 3s primrose and 3s 11d geranium, being clipped for either single or return journeys, the printing being black on the coloured boards as listed, with the whole order totalling £11 11s 8d.

Also notable for the Autumn Term of 1957 was the increase to 40 Scholar's Season Tickets ordered by Berkshire CC Education Authority, as opposed to just 21 the previous academic year, this being the time when the post-war 'baby-boomers' were coming of age for transfer to secondary school.

Indeed, the relative lack of such places in the area, especially when single-sex or mixed schools in an system them part Grammar, Secondary Modern and Denominational, would soon lead to a vast increase in the bussing of pupils, such relatively limited needs up until then usually being met by service buses, even if 'reliefs' were required. The squeeze on costs in the industry had also resulted in some reluctance by the larger operators to keep vehicles for limited use, so we shall see how the changes helped *White Bus* survive.

One local example was Leslie Grout, who lived at Mezel Hill Cottages in the Great Park, and was sent to Windsor County Grammar School for Boys, where the author remembers him from. He went on to win the coverted 'Mastermind' trophy in 1982 at Christchurch in New Zealand, the first time I'd seen him since I left the school! He stills lives locally and, as his father was Scoutmaster in the Park, so the *White Bus* featured throughout his earlier life in several ways.

The pressure on local schools resulted in the creation of a new Secondary Modern known as Charters School, built on a site by the junction of Devenish Road and Charters Road in Sunningdale, which from the outset was intended to draw from a wide area, but at first only had a roll-call of 400 pupils when opened by The Queen on 23rd April 1958. However, with the development of Bracknell New Town, and increases in population in Windsor and Ascot, it was expanded to 600 from 1970, then to 770 a year later, to 1000 in 1973 and later again to the present 1400, most of those additional children requiring transport.

The *White Bus* involvement with such operations started in a modest way, after Cecil was invited to put in a tender during March for a route starting out from North Ascot, though his rough notes of the time show there were 4 such routes being tendered in the area at that point, which became known as Coaches 1 to 4.

Also with education in mind, whilst all this was going on, young Douglas Jeatt had progressed to become a boarder at Salesian College in Farnborough, and in one letter home asked for copies of Practical Wireless and a specific valve to be brought over, whilst also noting he had seen the film 'Hornblower', seemingly both being a foretaste of his life in due course as a Radio Officer in the Merchant Service!

It is still notable that, although Cecil did not actively compete with his sibling, quite a few offers of private hires came his way, but when a 38-seater was called for in July 1958 for a trip from Sunninghill to Southsea he hired in one from Windsor-based *A. Moore & Sons (Imperial)*, another long-established local operator he was on good terms with. On the services front, the opportunity was taken when the license for the service through the Park was renewed in July 1958, to add a provision for extra late journeys to run after dances or sporting events.

With the extra numbers of pupils now travelling on the Park service, a larger vehicle was sought, and in October 1958 a front-engine Albion 'Victor' FT39N-type (OPG 991) was acquired. New in 1950 it was originally with *Green Luxury Coaches* of Walton-on-Thames in Surrey and carried a fully-fronted 31-seater front-entrance coach body by Allweather. Powered by a 4.88-litre diesel engine by Albion it was often found covering the bus routes alongside the 'Comet'.

Above- Photos of the Albion 'Victor' (OPG 991) in service are rare, but John Aldridge captured it at the iconic Copper Horse in October 1959, and it continued to wear the green livery of its original owner. *Below-* Both the 'Victor' and 'Comet' (KXU 673) were withdrawn in January 1960, the Albion as a result of a crankshaft failure. They remained in the yard for some time whilst their possible return to service was being considered, but were then sold during 1961. This photo was probably taken by John Gillham, who paid several visits there in 1960/1.

1959-61 – Into The 'Sixties

Although the fleet had suffered from a number of collisions out on the public highways over the years, an accident of a different nature saw one of the buses damaging the gates at Ranger's Gate on 17th February 1959, which resulted in a bill from the Commissioners for Crown Lands of £9 3s 3d!

Apart from that incident the first half of that year was a quiet one for changes at *White Bus*, but from June the swimming journeys also saw the Royal School in the Great Park and St. Catherine's School in Windsor added to those covered, though the pool used was not noted as such.

Private hires also increased, with the regular Ice Hockey at Wembley, plus a visit of the Moscow State Circus to that same venue catered for. Also noted in 1959 was the first conference for the Law Society at

Cumberland Lodge, which would become a regular booking for many years. The Sunningdale Football Club and Ascot Gas Cricket Club used *White Bus* for their away fixtures, whilst several Brownie packs also did for outings or to camps. Another regular Saturday job was for the Windsor & Eton Athletic Club to away events, taking a vehicle to Binfield (£3), Farncombe, Guildford, Hayes, Wycombe, Maidenhead, Maidstone (£9 9s), Pangbourne, Southall, Uxbridge and Watford.

On one occasion in June a hire for 4 coaches to Wallingford Police Sports resulted in the hiring of 3 vehicles from *Peggies Coaches* of Tilehurst, whilst another work ticket reveals that a Commer 'Commando' (GOM 658) was hired from *Winkfield Coaches* in May 1959, plus an exceptional hire to Littlehampton for a annual work's outing in August was covered by *Imperial* by a 41-seater and a pair of 38-seaters. It seems that Cecil also had to concede that covering the morning school contract was too much, so it was sub-let to *Imperial* from the Autumn Term 1959, which continued for some years.

In the meantime the former *Borough Bus* Bedford OWB (CMO 495) had been withdrawn in August 1959, though not replaced as such at that time.

Generally speaking the *White Bus* fleet was a well-maintained one, but reliance on secondhand purchases and outside parking, along with the poor quality timber used in many post-war bodies, could cause problems. On one occasion in October 1959 the PSV Examiner pulled a spot inspection on Bedford OB MVX 508 at Ascot. He found that the overhead locker door was decayed and several seats had worked loose, so he imposed a 'stop notice' until rectified.

Other issues of officialdom were also of concern to both *White Bus Services* and *Winkfield Coaches* in November 1959, after they submitted a joint planning application for a new bus and coach shed. The plan had called for the demolition of the old shed that had developed from the 1930's, and replacement by a new covered, but open-fronted, garage building. As the local planning authority, the Easthampstead RDC wanted to see sliding doors on the front of the garage, whilst Berkshire CC, as highway authority, was concerned that the site and proposals were not suited to the village setting, so it advised on a close-boarded fence or screen-planting to hide it from the road. As it was, the application was rejected and, after an appeal, was again rejected, so nothing changed on site.

For the second half of the 1950's the number of cars on the road took a notable upward turn, with makers aiming production at economical family cars within the budget of the masses. All bus companies large and small saw their takings reduced, and at a time when rising costs for the industry including the fuel tax, were starting to bite harder. On the Bagshot service in

November 1959 the income amounted to just £291 9s 3d for the month, with one Sunday netting just £3 14s 2d, whilst the best Saturday brought in £13 16s 2d. Fortunately the next month saw some useful bookings to see the Christmas Lights in London's West End at 6s 6d per adult and 3s for children.

January 1960 saw the withdrawal of the Albion 'Victor' (OPG 991) after a crankshaft failure, whilst that same month the Leyland 'Comet' (KXU 673) was also laid up, with neither vehicle returning to service.

Leyland 'Tiger' JXH 661 at Windsor Central Station.

As a result of those withdrawals another Leyland was acquired during January 1960, though quite unlike the bonneted 'Comet', in the shape of an oil-engined Leyland 'Tiger PS1/1-type, carrying a 33-seater front-entrance coach body by Thomas Harrington of Hove. Very much a typical coach of its size for its date new of 1948, still largely to a pre-war design, it was one of the large Stamford Hill-based fleet of *George Ewer & Co.* of North London, running under the main fleet of *Grey-Green Coaches,* which also describes the livery it continued to carry. As that firm ran various express coach services from London to East Anglia and the Kent Coast, it was in the habit of painting such place names on the louvres over the side windows, and close examination of photos of this coach reveals that 'Margate' could still be seen above the entrance door!

It should be noted that no fleet numbers were used at *White Bus,* but at 1960 the work tickets show that the registration numbers were referred to, despite not having two vehicles with the same letter mark in use.

Many of the private hires for 1960 followed a similar pattern to the previous year, with the variety shows at the London Palladium very popular, and not yet televised, which would ultimately slash seat bookings affecting the theatre and coach operators alike, another facet of social change then emerging. A new venture that year saw a coach travelling in May to the bulb-fields around Spalding in Lincolnshire, the 30-seater covering the round trip in a day for £26 14s, the longest single trip so far recorded!

John Gillham found the former Grey-Green 'Tiger' working the Windsor to Bagshot service in August 1960, and here we see it at The Square in Bagshot. In this photo are 3 cars, whereas when Bedford WTB (BUN 677) was captured a decade before there were none in that same view. A painted, rather than chrome plated radiator shell highlights post-war shortages.

June 1960 saw yet another chassis make added to the fleet, an AEC – or to be more correct an AEC rebuild by the Rochester-based coachbuilder Beadle. That firm had extended the lives of many bodies during the war years, and developed ideas to alleviate the post-war shortage of new chassis by incorporating older units refurbished from out-moded types into 'new' semi-chassisless vehicles, recycling in action in fact!

So those who subsequently rode on (or even drove) the example coming to *White Bus* (NKT 934) would hardly suspect that it started life in 1935 as an AEC 'Regent' double-decker ordered by *Autocar* based at Tunbridge Wells in Kent, though actually delivered to *Maidstone & District* as DH306 (CKE 440) after they acquired that firm. It ran in that form until early 1951 when it was withdrawn and the 7.7-litre oil engine and other mechanical parts were removed and rebuilt by Beadle into a 35-seater front-entrance coach for further service back with *M&D*, finishing off its days on bus work. By the time it reached Berkshire it had coach seating again, and was also popular for having a radio, so it was used for excursions. Its livery with *White Bus* has not been confirmed, but the cream with dark green trim of the Kent operator seems likely to have still been carried.

Discussion of this vehicle also brings us to some interesting points regarding coach operations in that era, before the days of mobile phones or internet access. On one occasion that vehicle was on a trip to London's Theatreland when, having just dropped off the passengers near Shaftsbury Avenue, the foot-brake pedal became stuck in the down position. A call to the office from a coin-operated phone box resulted in another to the AEC Service Depot, not far away at Southall in Middlesex, with the immediate dispatch of a fitter and van. The fault was soon rectified before the performance of the show ended, so all was well! In those days a box of business cards was as essential as our modern aids, and it is notable that Cecil's box also contained many for many other coach operators 'just in case' for breakdowns or possible hires.

The Beadle-AEC (NKT 934) is seen at North Street in this photo by John Gillham dated October 1960, the front-end treatment being typical of such vehicles. To its right is the rear of the Leyland 'Comet' (KXU 673) and its classic Windover rear end arrangement with orange 'flashers' each side of the registration plate.

This nearside view of NKT 934 also includes a young Paul Simons, a schoolboy whose parents had the Bon Marche clothes shop in Sunninghill. He liked to visit the yard and help clean the buses, but did not join the staff, becoming a Jesuit priest instead.

Berkshire CC Education Authority requested swimming transport from 16th May 1960, which saw the children from the Royal School in the Great Park and those from St. Francis RC School in South Ascot going to the pool at Little Paddocks. That swimming pool was in the grounds of a school for partially-sighted girls based in a Queen Anne mansion once the home of Colonel Horlick, the founder of the malted hot drink of the same name and manufactured over at Slough. The school was off the London Road at Sunninghill, now incorporated in the luxury Royal Berkshire Hotel, though *White Bus* only charged £1 for the trips from the Park and £1 5s for those from St. Francis's.

An example of a 1960 Scholar's Season Ticket, which is shown here a little larger than its actual 2.5 inch size. These were printed on pink card, and note the option for a 5 or 6-day validity, some schools having attendance for pupils on Saturday mornings.

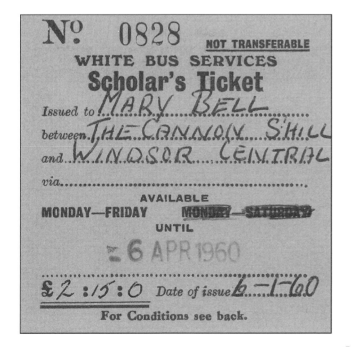

By September 1960 there is confirmation that *White Bus* was responsible for Coach 3 to Charters School, with pick up points by J&T Engineering around the mid point of New Road and another in Fernbank Road by Ascot Heath School, both points being in North Ascot. The numbers had risen from 36 in 1959 to 42, and of course in those pre-seatbelt days the generally accepted rule was that 3 children could occupy a seat intended for 2 adults.

Some mention was made in a photo caption earlier to the visits to the North Street yard by the well-known transport enthusiast John Gillham, who particularly liked to investigate the smaller independent operators on the fringes of the London area. After such visits to Windsor operators during 1960 he also spoke to Cecil Jeatt, and his subsequent article in 'Transport World' recorded some interesting facts.

He noted that *White Bus* was operating a school journey in supplement to the normal bus route running via Blacknest Gate – Cumberland Lodge – Queen Anne's Gate and into Windsor, adding that Cecil Jeatt and one full-time driver maintained the services during the daytime, with 2 part-timers covering the evening work. Once again it was mentioned that any private hire work was not actively sought, in respect of the division with *Winkfield Coaches,* but was not turned down

As to income from the services, he went onto to say 'the area served is sparsely populated, and so many former passengers now have cars, that the bus services scarcely cover expenses, and could hardly survive without the valuable help of Mrs. Jeatt, a school teacher. Mr. Jeatt is fully dedicated to the business and does not want to give up'. He adds 'the fleet was formerly entirely Dennis, but owing to falling traffic new buses cannot be afforded, so he has to buy second hand ones, though fortunately he has the ability to keep the existing fleet in tip-top mechanical order'.

In a final comment John reminds readers 'that Mr. Jeatt is typical of many stalwart pioneers who built the British bus industry into what it is today, and one cannot help admiring him for carrying on so valiantly in the face of great difficulty'.

He also quotes from *Bert Cole (Blue Bus Service)* of Slough, who decries the burden of fuel tax on bus operators, who also adds that it must either be reduced for rural services, or other forms of subsidy found if such routes were to survive.

1961 saw quite a few changes in the fleet, though not everything went to plan. Any prospect of the 'Comet' (KXU 673) returning to service had by then faded, as Cecil did not have the energy to rectify its issues, so early that year Mrs. Jeatt bought a coach from a Chertsey dealer, which proved to be a mistake.

The vehicle concerned was a Morris OPR-type (KOE 207), one of a batch built for *Blue Cars Continental Tours* of London in 1949 and fitted with Plaxton 22-seater front-entrance coach bodies. That operator did indeed run tours across the Channel, though in those days the average British coach body design was more concerned with maximising bums-on-seats and being water-tight than its continental counterparts, which generally had more spacious seating, better all-round visibility and air-conditioning!

The Morris coach (KOE 207) with its rebuilt body incorporating continental styling including the marker lights fitted at roof level, the radio aerial and the fin on the front dome in this photo by Roy Marshall.

For 1950 *Blue Cars* had a dozen underfloor-engine Leyland 'Royal Tigers' on order with bodies by Beccols of Chequerpoint, near Bolton, Lancashire, so they sent the Morris coaches for some modifications, in view of the fact that Beccols had won an award at the Montreaux International Coach Rally in 1950.

Although this would seem to be highlighting a good end to this story, once the body was further checked over at Winkfield, the extent of deterioration became apparent, so although a start had been made on the mechanical side, the project had to be abandoned!

This John Gillham photo shows the yard in June 1961, with the rear offside of the Morris coach, complete with 'GB' plate, alongside the 'Comet' and 'Tiger'.

Bedford OB (JXH 719) at Windsor Central Station as caught by John Gillham in April 1962 on a layover.

A rather safer bet followed, with the purchase of another Bedford OB with Duple 'Vista' 29-seater front-entrance coach body (JXH 719) in May 1961, new in April 1948 to the *George Ewer* subsidiary of *Fallowfield & Britten* of London N16. After sale it had several owners, the one before *White Bus* being *Winkfield Coaches*, which had acquired it in June 1956, and it now meant two vehicles bearing the same registration letters once again featured in the fleet.

It should of course be noted that most of these coach-bodied vehicles featured sliding entrance doors, not ideal for use on service routes with a driver-conductor, though a widespread practice at the time. In order to overcome that a system of levers was fitted under the floor so that the driver could open and close the door from a seated position, not necessarily an easy task at times! In respect of the 'Vista' bodies, the coachbuilder Thurgood of Ware had come up with a patented design, a copy of which has been found in the *White Bus* archive and could be built from 'scrap'.

May 1961 saw an unusual amount of private hire jobs being covered by *Imperial,* suggesting that Cecil may have been in hospital, and indeed that year it was with one of its coaches that the Spalding Bulb Fields trip was covered, as well as local jobs for the Fleur-de-Lis pub at Lovel Hill. By then Rod Haylor was appearing on work tickets as a part-time driver, his daytime job being with the forestry department in the Great Park.

A number of detailed revisions to services were applied from 1st July 1961, obviously aimed at reducing wasted mileage or to improve loadings. On the Bagshot service the last run from Windsor now ran only as far as The Crispin on Mondays to Fridays, extended onto Sunninghill only on Saturdays and Sundays, before running back dead to its base. Some Windsor – Sunninghill journeys now omitted the loop through Woodside, though some of those onwards from Sunninghill did still pass through Windlesham village. On Saturdays there was a trip to Sunninghill which ran past Cumberland Lodge, useful for shopping or an afternoon at the Picture House.

On the Windsor – The Crispin route a schooldays run reflected the growing number of scholars travelling from the Park into Windsor, with the 8.7am departure from Cumberland Lodge running via Queen Anne's Gate to reach Windsor Central Station, having passed the stops for Brigidine Convent School in Kings Road, then Osborne Avenue for the County Girl's School and Trevelyan School a short walk away in St. Leonards Road, past Victoria Barracks to drop off for Royal Free in Batchelor's Acre, to the Station where the County Boy's walked to Maidenhead Road and nearby in Vansittart Road the girls of Princess Margaret Rose. In the afternoon each school turned out in time for the children to reach the Station for the return trip or board on the route as it worked out of Windsor. On the downside the 7pm journey from Windsor to The Crispin now ran only on Saturdays.

Also included in the 1961 application was a general increase in fares, another subject which obviously gave Cecil a lot of mental exercise, and these became-

From	To	Single	Return
Windsor	Bagshot	2s 4d	4s 0d
Windsor	Sunninghill	2s 9d	3s 0d
Windsor	The Crispin	1s 0d	1s 9d
Windsor	Cumberland Lodge	11d	1s 8d

As noted during the previous year, the Bedford OB (MVX 508) was suffering in the body department, so in November 1961 it was replaced with another very similar 1950 model (TMY 950), which had originated with *Thorne* of London SW2.

Incoming Bedford OB (TMY 950) is seen here in this April 1962 photo by John Gillham at Windsor Central Station awaiting departure on a service run. Despite the long scrape evident in this view, it proved to be an enduring vehicle and lasted through to 1966.

1962-3 – To Bagshot No More

We have seen how part-time drivers helped keep all commitments covered, and during 1960-2 there are numerous references to Ray Bartrop, and like other men so employed he also had a daytime job. He also organised seaside trips from his South Ascot home area under the guise of 'the Gay Gadabouts'.

For April 1962 we can review a breakdown of the scholar's passes paid for by Berkshire CC Education Department, with 19 for the County Boy's, 14 for the County Girl's, 5 for Princess Margaret Rose, 4 for Royal Free and 11 to Trevelyan, all using the route from Sunninghill via Blacknest and Cumberland Gate, whilst there were also girls for Brigidine Convent not so funded as it was a private establishment.

During June 1962 the former *Grey-Green* Leyland 'Tiger' (JXH 661) was withdrawn, and it would remain the only half-cab vehicle ever owned by the firm, whilst its 7.4-litre engine drank too much fuel compared with the lightweight Bedfords.

In its place from October in time for the Autumn Term 1962 was another Albion 'Victor' FT39N with Allweather 31-seater front entrance coach body new in 1950 to *Green Luxury Coaches* of Walton-on-Thames. OPB 750 originally had a petrol engine, replaced by the diesel Albion version in March 1958.

This second Albion was practically identical to the other example at White Bus, but they were never in service at the same time. Most 'Victors' of the period received fully-fronted bodywork, and once again the original green livery was retained, though this photo does show it with the first owner.

The issue of the garage had not gone away and, whilst it is not clear what happened after the refusal for the new structure, in July 1962 a letter came from ERDC ordering demolition of the unauthorised structure! The Jeatts and Maulers put in an appeal, which was lodged too late for the prescribed deadline. However, some form of compromise was evidently struck, as the site remains as open to the road frontage as it ever was.

By the Summer of 1962 it is clear that some serious soul-searching was being done to find a way that the bus services could continue, the surviving workings in rough by Cecil showing how it exercised his mind.

There was great loyalty from *White Bus* to its users and, despite the increase in car ownership, many still relied on the bus, as within a household the husband might buy a car, but his wife needed to go to the shops whilst he was out working, and the children had to get to school, many motorists then being 'week-enders'.

Economy of operation was needed, so the core issues of serving the communities the Park and those at points not served by the buses of other operators. It should also be noted than in the correspondence with the Traffic Commissioner, the latter displayed a good sense of trying to help maintain such links, thereby dispelling any idea of unsympathetic officialdom.

So, instead of a standard renewal application in July 1962, the operations were re-arranged in order to make them into one service, though incorporating a variety of journeys. The main provisions were to abandon the section of route from Sunninghill through to Bagshot, to end all Sunday operations, and to merge the two separate routes onto one timetable in time for changes in October.

It should be appreciated that those passengers along the A30 towards Windlesham and Bagshot had the buses of *Aldershot & District* to ride, whilst various parts of the Sunninghill and Sunningdale areas were served by the *Thames Valley* 'Ascot Local Services' Routes 2a/2c, which connected with the mainline Route 2 (Windsor – Ascot – Reading) at Ascot High Street and Southern Region trains at Ascot Station. The proposals also safeguarded the journeys used by schoolchildren from Sunninghill, Cheapside, and the Park for their travel to the Windsor schools.

This produced a very complex pattern of travel over the week, with some journeys through the Great Park, others running more directly, some scheduled relief workings or school-day runs, so in order to do justice to that the full timetable appears on page 63.

The final operation of the separate service to The Crispin ran on Saturday 27th October, whilst both the Bagshot portion and Sunday operations in general ended the following day, with the new timetable effective from Monday 29th.

It is worth noting that subsequent letters from the Traffic Commissioner dispensed with the need to make a formal application to try this 'experimental' timetable for the Winter period 1962/3, which seems to be further evidence that the Commissioner realised it was an issue of whether the service could survive.

By March 1963 it was evident that the arrangements were working, so *White Bus* was asked to surrender the old versions of the separate licenses at the next renewal date, whilst the Commissioner wrote again in August to enquire as to how the service was faring.

The only vehicle change during the year was the withdrawal in June 1963 of Bedford OB (JXH 719), after only 2 years of use. It is worth noting that it and the 'Tiger' had both been bodied in 1948, not a good year for timber for coach-building in general. However the Bedford did not depart for some years, slowly descending into the undergrowth at the rear of the yard, where the author encountered it on his visit!

1964-1966 – Bedford Interlude

Around the time of the curtailment of the route at Sunninghill, the Clancy family moved to nearby the High Street there, and Michael recalls that the bus then turned into School Road to travel back to the terminal point ready to return to Windsor.

During March 1964 a larger Bedford was acquired from *North Star Coaches* of Stevenage, Hertfordshire (SAR 128), a 1954 SBG-type with 4.92-litre Bedford petrol engine and 18ft wheelbase. It carried a Duple 'Vega' 36-seat, front-entrance coach body, recalled for its nice interior finish, whilst externally it retained its original grey and maroon livery with *White Bus*. It became a regular performer on the bus service, and Michael Clancy recalls often riding on it to Windsor and back.

Bedford SBG (SAR 128) had the oval style of radiator aperture, and it was caught by the camera of Philip Wallis taking on passengers at the White Bus stand in Windsor Central Station in February 1967.

The SBG ousted the Beadle-AEC (NKT 934), which has a departure date of July 1964, leaving the Albion 'Victor' as the only non-Bedford chassis in the fleet. Incidentally, nearby in New Road, North Ascot J.T. Engineering had acquired a very similar-looking bus for staff transport, being a Beadle-Leyland (OKP 984) which was also recycled from within the *Maidstone & District* fleet but using Leyland running parts.

	School Relief a.m.	School Term a.m.	No School a.m.	A a.m.	B a.m.	Th B a.m.	A p.m.	B p.m.	p.m.	SR p.m.	p.m.	p.m.
SUNNINGHILL SCHOOL dep.	7.55	8.05	7.55	9.22	9.22	11.32	1.32	1.32	3.32	4.33	4.42	5.42
Cannon	7.58	8.08	7.58	9.25	9.25	11.35	1.35	1.35	3.35	4.36	4.45	5.45
Blacknest Gate	8.01	8.11	8.01	9.28	9.28	—	1.38	1.38	3.38	—	4.48	5.48
Cheapside (Post Office)	—	8.15	8.05	9.32	9.32	11.39	1.42	1.42	3.42	4.40	4.52	5.52
Sawyers Gate	—	8.17	8.07	9.34	9.34	11.41	1.44	1.44	3.44	4.42	4.54	5.54
Cumberland Lodge	8.10	—	—	—	9.38	—	—	1.48	—	—	—	—
Crispin Hotel	—	8.20	8.10	9.37	—	11.44	1.47	—	3.47	4.45	4.57	5.57
Woodside	—	8.22	8.12	9.39	—	—	1.49	—	—	—	—	—
Royal Lodge	—	—	—	—	9.40	—	—	1.50	—	—	—	—
Forest Gate	—	8.26	8.16	9.43	—	11.47	1.53	—	3.50	—	5.00	6.00
Village (Post Office)	8.14	—	—	—	9.44	—	—	1.54	—	—	—	—
Wheeler's Yard	8.16	—	8.19	9.46	9.46	—	—	1.56	—	—	—	—
Queen Anne Gate	8.20	8.31	8.23	9.50	9.50	11.52	1.58	2.00	3.55	—	5.05	6.05
WINDSOR STATION arr.	8.25	8.36	8.28	9.55	9.55	11.57	2.03	2.05	4.00	—	5.10	6.10

	SR a.m.	a.m.	Th a.m.	B a.m.	p.m.	A p.m.	B p.m.	SR p.m.	p.m.	p.m.	p.m.
WINDSOR STATION dep.	8.30	8.30	10.55	10.55	12.05	2.55	2.55	4.05	4.05	5.15	6.15
Queen Annes Gate	8.35	8.35	11.00	11.00	12.10	3.00	3.00	4.10	4.10	5.20	6.20
Wheeler's Yard	—	—	11.04	11.04	12.14	3.04	3.04	4.14	4.14	5.24	6.24
Village (Post Office)	—	—	—	11.06	—	—	3.06	4.16	—	—	—
Forest Gate	8.40	8.40	11.07	—	12.17	3.07	—	—	4.17	5.27	6.27
Royal Lodge	—	—	—	11.10	—	—	3.10	—	—	—	—
Woodside	—	—	11.11	—	12.21	3.11	—	4.21	—	—	6.31
Crispin Hotel	—	—	11.13	—	12.23	3.13	—	4.23	—	5.30	6.33
Cumberland Lodge	—	—	—	11.12	—	—	3.12	4.20	—	—	—
Sawyers Gate	8.45	8.45	11.16	11.16	12.26	3.16	3.16	—	4.26	5.33	6.36
Cheapside (Post Office)	8.47	8.47	11.18	11.18	12.28	3.18	3.18	—	4.28	5.35	6.38
Blacknest Gate	8.51	8.51	—	11.22	—	—	—	4.27	4.32	—	6.42
Cannon	8.54	8.54	11.22	11.22	12.35	3.22	3.22	4.30	4.35	5.39	6.45
SUNNINGHILL SCHOOL arr.	8.57	8.57	11.25	11.25	12.38	3.25	3.25	4.33	4.38	5.42	6.48

A—Mon., Wed., Thurs.   B—Tues., Fris.   Th—Thursday only.   SR—School Relief only.

	a.m.	a.m.	a.m.	a.m.	p.m.	p.m.	p.m.	p.m.	p.m.	p.m.	p.m.	p.m.	p.m
SUNNINGHILL SCHOOL dep.	7.55	9.22	—	11.32	1.32	—	3.02	3.32	4.32	5.32	7.32	9.32	10.52
Cannon	7.58	9.25	—	11.35	1.35	—	3.05	3.35	4.35	5.35	7.35	9.35	10.55
Blacknest Gate	8.01	9.28	—	—	1.38	Rel	—	3.38	4.38	5.38	7.38	9.38	—
Cheapside (Post Office)	8.05	9.32	—	11.39	1.42	1.42	3.09	3.42	4.42	5.42	7.42	9.42	10.59
Sawyers Gate	8.07	9.34	—	11.41	1.44	1.44	3.11	3.44	4.44	5.44	7.44	9.44	11.01
Cumberland Lodge	—	9.38	—	—	—	1.48	—	3.48	—	—	—	—	—
Crispin Hotel	8.10	—	10.24	11.44	1.47	—	3.14	—	4.47	5.47	7.47	9.47	11.04
Woodside	8.12	—	—	—	1.49	—	3.16	—	4.49	5.49	7.49	9.49	
Royal Lodge	—	9.40	—	—	—	1.50	—	3.50	—	—	—	—	
Forest Gate	8.16	—	10.27	11.47	1.53	—	3.20	—	4.53	5.53	7.53	9.53	
Village (Post Office)	—	9.44	—	—	—	1.54	—	3.54	—	—	—	—	
Wheelers Yard	8.19	9.46	—	—	1.56	—	3.56	—	—	—	—	—	
Queen Annes Gate	8.23	9.50	10.32	11.52	1.58	2.00	3.25	4.00	4.58	5.58	7.58	9.58	
WINDSOR STATION arr.	8.28	9.55	10.37	11.57	2.03	2.05	3.30	4.05	5.03	6.03	8.03	10.03	

	a.m.	a.m.	a.m.	p.m.	p.m.	p.m.	p.m.	p.m.	p.m.	p.m.	p.m.	p.m.
WINDSOR STATION arr.	8.35	10.05	10.55	12.05	2.15	2.55	3.55	4.25	5.15	6.15	8.15	10.15
Queen Annes Gate	8.40	10.10	11.00	12.10	2.20	3.00	4.00	4.30	5.20	6.20	8.20	10.20
Wheeler's Yard	—	10.14	11.04	12.14	2.24	3.04	4.04	4.34	5.24	6.24	8.24	10.24
Village (Post Office)	—	—	11.06	12.16	—	3.06	—	4.36	—	6.26	—	—
Forest Gate	8.45	10.17	—	12.21	2.27	—	4.07	—	5.27	6.31	8.27	10.27
Royal Lodge	—	—	11.10	—	—	3.10	—	4.40	—	—	—	—
Woodside	—	10.21	—	12.25	2.31	—	4.11	—	5.31	6.35	8.31	10.31
Crispin Hotel	8.47	10.23	—	12.27	2.33	—	4.13	—	5.33	6.37	8.33	10.33
Cumberland Lodge	—	—	11.12	—	—	3.12	—	4.42	—	—	—	—
Sawyers Gate	8.50	—	11.16	12.30	2.36	3.16	4.16	4.46	5.36	6.40	8.36	10.36
Cheapside (Post Office)	8.52	—	11.18	12.32	2.38	3.18	4.18	4.48	5.38	6.42	8.38	10.38
Blacknest Gate	8.56	—	11.22	12.36	2.42	3.22	—	4.52	—	6.46	8.42	10.42
Cannon	8.59	—	11.25	12.39	2.45	3.25	4.22	4.55	5.42	6.49	8.45	10.45
SUNNINGHILL SCHOOL arr.	9.02	—	11.28	12.42	2.48	3.28	4.25	4.58	5.45	6.52	8.48	10.48

The combined and revised timetable introduced from October 1962, showing the full weekly times and routes.

Although Sunday operations had ceased, the special runs for the polo season between Windsor (Central Station) and Smith's Lawn had continued in 1963, and for 1964 they re-commenced from Whit Sunday 17th May. The local cinemas had been suffering from the national decline in audiences as more people bought a television set, so they now resorted to Bingo to fill the auditorium, and from May 1964 we see the first of the regular coach hires to take devotees into Windsor from the York Club in the Great Park. There were dances at the Royal British Legion Hall at Hatchet Lane in Cranbourne, for which *White Bus* ran a late journey. Also regular hires by St. Edward's School of Dorset Road in Windsor, though the venues varied from week to week, each taking place on a Friday, that school then having Mrs. Jeatt as a teacher!

St. Francis RC School in South Ascot also featured in regular hires to the London museums, whilst other schools covered that year were Sunningdale to the Bracknell Stadium and Priestwood School took a river trip on the Thames in London. The Windsor-Goslar Youth Exchange visited the Tower of London, the Brylcream factory near Maidenhead, and also went to Bognor Regis and Chessington Zoo.

It is notable that quite a few jobs were marked 'WC', so evidently covered by the Maulers, including the last trio mentioned above. On the other hand, *White Bus* loaned the Bedford SBG (SAR 128) and Mike Groves to *Winkfield Coaches* when a hire called for its 36-seat capacity, so ultimately the work got shared out between the two concerns to generally mutual benefit.

That the combined scheduling had now settled down is confirmed by the cancellation of the old Road Service Licenses for the former separate routes from July 1964, after a few minor changes in the light of experience since the rather severe Winter of 1962/3.

Both 1965/6 were years of very little of note for *White Bus,* other than some fleet changes. The Albion 'Victor' (OPB 750) was withdrawn at the end of April 1965, being replaced by another Bedford SBG (LEA 500). However, unlike the previous example, this carried a Burlingham 'Baby Seagull' body with front entrance and 36 coach seats, being new in 1955 to *Hill* of West Bromwich, though it came to Berkshire via several others.

The incoming Bedford SBG (LEA 500) was caught by Philip Wallis at the North Street yard in February 1967. Note the withdrawn Bedford OB (JXH 719) on the left, laid up back in June 1963 but still evident.

The departure of the Albion now rendered the fleet as all-Bedford for the first time, a situation that carried on until October 1967. The only fleet change of 1966 was the sale in February of Bedford OB (TMY 950), which only went as far as Windsor County Grammar School for Boys, which took it for a trip to the Continent, though after the author had left the school.

1967-9 – The Summers of Love

1967 heralded 'The Summer of Love', when Britain experienced the coming of the hippie culture and West Coast music influences, followed by the notable series of music festivals on the Isle of Wight. However, at *White Bus* it was a remarkably uneventful year, with the bus service settled down and other commitments very similar to before.

Indeed, there was just one fleet change for 1967, with a break in the Bedford monopoly by the arrival of a Ford 570E-type chassis (YFH 53), which carried a 41-seater Plaxton 'Consort IV' front-entrance coach body new in 1959. It came from *Neale* of Teddington in Surrey, and represented a more active effort by Ford to re-enter the lightweight PSV market.

This view of Ford 570E coach (YFH 53) shows it on a layover by the Sunninghill Picture House with driver Jim Collins, who started part-time during 1967.

A second contract run to Charters School commenced from February 1967, and *White Bus* was now covering Coach 3 (Hatchet Lane – Brookside – Kennel Ride – J.T. Engineering, New Road – Ascot Heath School, Fernbank Road – Goaters Lane, Fernbank Road), with 50 children on pink season tickets, and Coach 4 with 45 pupils on green passes (Prince Albert, Clewer Green – Fernbank Road Corner). *Moore's (Imperial)* continued to cover one morning journey, that being the one emanating close to their base in Clewer Hill, as well as helping out with some private hire jobs.

This offside view of the Ford (YFH 53) is courtesy of Phil Moth and shows it opposite the Parish Church in Windsor, having just arrived in the town on service.

Despite the increase in daily commitments, one of the Bedford SBG's (LEA 500) was withdrawn and sold after a short stay in March 1968, whilst the other of that type (SAR 128) was withdrawn in June 1968. In respect of the latter the Certificate of Fitness expired.

However, it had been the intention to return it to service in due course, its rather nice interior having already been noted. As time went by such work slipped due to other more pressing issues that duly turned to a hope that it might go for preservation, but sadly it finally went for scrap after 1972.

The incoming vehicle to replace the withdrawals was another completely different departure, and the first underfloor-engine type in the fleet. Although it was in fact the third Albion owned, it was of a quite scarce 'Aberdonian' MR11N-type, which had been produced as a lightweight rival to the Leyland 'Tiger Cub' (CU 9756). It featured a horizontally-mounted Leyland 0.350 engine of 5.76-litre capacity and a 5-speed gearbox. New in 1957 it was part of a batch delivered to *Northern General* in Tyneside for one-man use with its Weymann 44-seater front-entrance bus body and electronically-operated jack-knife door. Doug Jeatt recalls that it drove well, though the type was noted more generally for its poor braking!

There are no good photos of the 'Aberdonian' with White Bus, but nearby Moore's (Imperial) actually had its batch-mate (CU 9757), photographed by Tony Wright on a Local Service in Windsor in their brown and cream scheme. However, the example at Winkfield carried an all-over NGT dark red livery.

A little earlier, back in April, one of the longest serving employees first made his appearance, when former *Thames Valley* driver Jim Waterfall entered the ranks as a part-timer, though from January 1970 he went full-time, mostly on the service route. Around that same time Mick Fazey also started on a part-time basis, progressing onto full-time from March 1983.

As early as 1961 a Government-funded investigation, the Jack Report, had highlighted the need to provide subsidies for rural bus services in order that people in such areas would not be left without any transport links, a situation made more critical by the railway branch-line closures of the same era. At that juncture no action was taken, but the Transport Act 1968 did have provisions to address the issue. Under its provisions Local Authorities could make grants up to 50% to support rural services, though of course

Berkshire CC had rather a lot on its plate with most services in the west of that county then under threat.

A sign has graced the North Street yard from back in the 1930's, latterly on the fence of the Windsor side. Over the years the localised exchanges were absorbed into ever larger geographical areas, so here we see an interim stage, still 'Winkfield Row', with 88 added to the former 2612, but later under Bracknell's code.

From time to time *White Bus* was invited to consider some additional work, and one such occasion in June 1968 involved one in connection with the then popular wrestling matches held monthly at Bracknell Sports Centre on the junction of Bagshot Road and South Hill Road. The Recreational Facilities Manager had received representations that the *Thames Valley* bus service did not run at suitable times, so the Winkfield firm was invited to tender for a special run to carry ticket-holders for those events, but nothing came of it.

Amongst the varied paperwork of the *White Bus* archives are a number of letters from satisfied clients, mainly regarding special journeys, but a nice one arrived in September 1968 from a parent whose boys had used the buses for many years. Mrs. Ludlam of Silwood Farm, Cheapside sent her 'letter of grateful thanks to you and all your coach drivers, who have always shown such courtesy to me. Now that Hugo is old enough to use a motor-scooter he no longer needs to use your service to and from Windsor, and of course Richard ceased to travel by bus long ago. Both boys enjoyed their years commuting on your service, and I enjoyed the varied anecdotes they were able to recount to me on occasions! Thank you Mr. Jeatt, and Long Live the White Bus Service, a boon to so many inhabitants of this isolated area'. The author recalls Richard from Windsor Boy's, but in due course we shall hear of him once again in a more direct sense.

The most significant event of 1969 for bus operators was the changeover to a decimal currency from 1st August. A letter was circulated early in the year by the Traffic Commissioners to prepare operators, as old half-penny fares would need to be eliminated, and new structures cleared in good time, which acted as a form of control on profiteering, which was rumoured to be a consequence of the change. Out went the old pound, shillings and pence regime, replaced by a new pound and new pence, the 50p being the equivalent of the old 10 shillings (or half-a-pound), though few other calculations were quite as tidy as that.

The other notable event locally was the opening of the Windsor Safari Park by the Smarts, just a short way from their Circus Winter Quarters near the North Street yard, and carved out of Windsor Forest on St. Leonards Hill. The Smart family had long been circus people, and Billy Smart junior had taken over from his father in 1966, along with brothers Ronald and David. The circus ceased touring in 1971, but did regular television broadcasts both in Holland and the UK up to 1983. Guernsey Zoo was another venture, but sold off in 1972. Billy being a very eligible batchelor was linked at times to Diana Dors (who starred in a film set in the big-top), Jayne Mansfield and Shirley Bassey, but married a German-born Pan Am stewardess he met flying to Los Angeles in 1973.

The Safari Park, was mainly based on an open-air 'African Adventure', where people drove through in their cars, usually coming out short of a wing-mirror, hubcap or radio aerial, thanks to the free-roaming chimps! There were also aviaries and paddocks where other species were kept away from the lions, but the most enduring local scene was when elephants were walked down the road between the Winkfield quarters and the park site! The presence of that development did nothing for the *White Bus* service, and indeed the Safari Park found itself in financial trouble by 1977, after which it was sold for re-development into the huge money-making machine known as Legoland.

However, despite some fares increases of late, the *White Bus* bank statements only make pretty reading if you are partial to the colour red.

1970-1972 – Trying Times

In the meantime Douglas Jeatt went onto St. Joseph's College at Beulah Hill in North London for A-levels, then to Slough College where he studied engineering for a year before deciding it wasn't for him. It had been made clear by his parents he was not expected to join the firm if he wanted to work in another line, and he trained at Southampton College of Technology in 1967 before joining the Merchant Navy as a Radio Officer, travelling the oceans of the world.

However, when visiting home he did sometimes help with some mechanical work, and also trained for his PSV Driver's Badge, under the tuition of his Father or Uncle Dick, recalling lessons on the Maudslay 'Marathon' (SMU 212) and AEC 'Reliances' from the *Winkfield Coaches* fleet. His test was with the local Vehicle Examiner Mr. Plumridge, and Doug recalls he was a bit nervous, missing a turning and not the best of hill-starts up Thames Street towards the Castle!

But pass he did, with License No.KK57611 issued from 1st January 1970.

Another new type made its appearance in the *White Bus* fleet from May 1970, perhaps influenced by the above events? As noted, the Maulers had examples of the AEC 'Reliance' included since 1962, so the buying of former *A. Timpson & Sons Ltd.,* of Catford, London SE6 2MU3RV-type (VXP 508) was logical. It had been new in 1959 and featured a 7.685-litre AEC AH470 engine mounted under the floor amidships, 5-speed synchromesh gearbox and vacuum-assisted braking. Above the chassis was mounted a stylish 41-seater front-entrance Weymann 'Fanfare' body.

The Weymann-bodied AEC 'Reliance' (VXP 508) seen at the stand in Windsor Central Station. The original livery had been cream with maroon trim, but a log book entry shows that from January 1969 that became grey and green.

Back in March something of a local crisis in coach capacity had followed the announcement that both the *East Berks Services* and *Bowler's Coaches* were to cease for various reasons. With the latter the issue was that its base on some open land off Station Road in the town centre of Bracknell was being re-developed under the New Town, quite literally. However, after a time that operator and resume PSV work from a base in Sandhurst, then later on nearby at Owlsmoor.

The issue with *East Berks Services* was a different matter, that operation stemming originally from the *S.R. Gough (Gough's Bus Service),* which had in the late 1940's worked eastwards from Bracknell to not far short of Winkfield, though from the mid-1950's only coaching work had featured. In 1962 the family moved its garage business solely to a site in Warfield Street, a short way north of the town, selling the base at London Road to Peter Seed, who ran it as East Berks Garage, the coaches changing their fleetname at that point, though the Storeman (my Father Bert Lacey) stayed on, so I knew the small fleet and their drivers well! However, in March 1970 the coaching gave way to expansion as a Rootes Main Dealership.

Both of the above operators had contracts for the transport of schoolchildren on behalf of the County, so the latter was hard-pressed to find alternatives for the Summer Term, there being no other indigenous full-size PSV owners in Bracknell. For *White Bus* the contacts it already had with Preistwood and Binfield Schools marked it out as a candidate for additional contracts. Certainly, by September 1970 it was the contractor for Coaches 3 and 4 from New Road and Fernbank Road (assisted in the morning by *Imperial)*, whilst that operator had Coaches 6, 8 and 9, and Coach 5 from Cheapside was *Winkfield Coaches*. No details of Coaches 1, 2 and 7 are given, but maybe it was when *Windsorian* started running to the school.

Imperial also helped out with other practical issues, as it was to their yard in Clewer Hill Road that vehicles were taken for steam-cleaning before inspections. One consequence of the cessation of *East Berks* was that its former driver Ted Waterman, who lived at Chavey Down between Bracknell and Winkfield, now drove at times for *Winkfield Coaches* or *White Bus.* Ted was an experienced coach driver with a friendly nature, whilst as he did not mind early starts and weekend work, he often covered the day-long fishing trips to coastal destinations, along with other contract jobs.

Income from the bus service continued to be rather low, and correspondence shows that there was some reluctance by Cecil Jeatt to increase fares as much as should have been required, something which would later see those fares very low compared with the two larger operators in the area.

A further vehicle came into the fleet from November 1970, with a return to the Bedford SBG-type, though this example had by then got a diesel engine. It had originally run for *Smith's Luxury Coaches* of Reading (NRD 371) and was new in their distinctive orange and blue livery in 1957, carrying a 41-seater front-entrance Duple 'Vega' coach body. After leaving the Reading operator it had passed to *Martin* of West End and was repainted as pale blue and cream.

Bedford SBG (NRD 371) had the 'butterfly-style' radiator grille and is seen at Windsor Central Station, with a Thames Valley Bristol FLF6G behind.

Unfortunately the AEC 'Reliance' (VXP 508) had developed a fault, so it was out of use by March 1971. At that point in time the plan was to put it back on the road for further service in due course. By that same month the Albion 'Aberdonian' (CU 9756) was also out of use, but it was actually sold to *Imperial* in August, as apparently they made a good offer to use it as spares for its batch-mate.

This view of Timpson's AEC 'Reliance' (VXP 508) by R.H.G. Simpson shows it when on a trip to Oxford. Note the classis lines of the Weymann 'Fanfare' bodywork and the front door arrangement.

Figures are available for the bus service income for March 1971, when the whole month only netted just under £294, with the worst Monday as low as £4.63p, and the best Saturday at £23.84p. That compares with the £292 of 1959, but without the inflation of those 12 years! To make matters worse Cecil's health was also an issue as 1971 came to its close, so something needed to be done if the firm was to continue. Joe Sutcliffe, who was then a fitter with *Thames Valley* recalls that he spent some time helping *White Bus* keep going around that period, though in due course he was a full time driver for *Winkfield Coaches*.

Doug Jeatt decided that he needed to come home for a while to help his ailing Father out, returning from sea in the Summer of 1972. In the meantime he had been on the 'Patonga', a refrigerated cargo ship on the New Zealand, Australia and Montreal (MANZ) run, then the passenger liner 'Oronsay' on Australia runs with '£10 Poms', or cruising the Mediterranean and to Brazil. Back on cargo the 'Pando Point' saw him going to India via the Cape of Good Hope and back through the Panama Canal, all great experiences!

Over at Bracknell the void created by the relocation of *Bowler's Coaches* and the cessation of *East Berks Services* had resulted in two operators appearing. One was the taxi and car hire firm started by *Joe Cooper* back in 1947, using the cars given to him by Lord Downshire when the estate was sold and he lost his job. The other was a new venture, as Barry Brophy of Keates Green in Priestwood, Bracknell got his first coach as *Nuetown Coaches* in February 1972.

That period is also recalled for the emergence locally of the fungal tree infection commonly known as Dutch Elm Disease, which over the decade would devastate all the trees of that type within the Park, the passing of which removed many ancient rookery sites, as well as changing whole vistas.

There were few immediate changes at *White Bus* as the priority was to keep things ticking over. However, Doug did rebuild the engine of the withdrawn AEC 'Reliance' (VXP 508), though sadly that classic vehicle never did return to service. For a time it occupied the rear spot in the yard where Bedford OB JXH 719 had stood a decade before.

On 23rd September 1972 the first of the 'People's Free Festival' was held in the Great Park off Sheet Street Road, the organiser Ubi Dwyer claiming that the land was common-land until enclosed under King George III. Some 700 people turned up for the one-day event, but the following year it was repeated for the bank holiday weekend of 26-28th August with some 1400 attending. Although it was advertised again for 28th August to 1st September 1974, the 7000 or so crowd was too much for the authorities, who broke it up with violent action by the Police, 220 people being arrested and the organisers sent to prison. On the brighter side, Doug recalls that they made £300 at 10p-a-pop from the bus service, being a couple of miles from the two Windsor Stations and *Green Line* stops!

The issue of fares was again discussed for September 1972, particularly as they had lagged behind those for the same distances then charged by *London Country* and *Alder Valley*, both operators supporting the *White Bus* submission for increases.

The Yeates 'Pegasus' bodied Bedford SB5 (6881 R) on the stand at Windsor Central Station in a photo by Tony Wright.

In respect of the fleet there was a need to buy something suitable for the service route, which could at other times be used on hires. A 1963 Bedford SB5-type was obtained, being one of a number altered by the Bedford dealer and coachbuilder W.S. Yeates of Loughborough, which set the front axle back in order

to accommodate a door opposite the driver for one-man-operation. That resulted in a near-vertical steering column, but gave room for 45 coach seats, though the 5.42-ltre diesel engine was still front mounted. Bodies from that maker were not common in Berkshire, and that example (6881 R) was new to *Frost* of Stanley in Derbyshire. It had subsequently been sold to *Morley* based in Cambridgeshire, who ran into Peterborough as the *Whittlesey Bus Service,* and it seems that when the vehicle was viewed in September 1972 the fact that it bore 'WBS' in various places influenced the decision to obtain it, retaining the red and cream livery it came with!

Bus-grant Bedford YRQ-type (LJB 403L) is seen turning in the yard of Windsor Central Station on a trip out to The Village in a photo by Phil Moth. Note the yellow ochre skirt, which first featured on it.

Mention has also be made of the funds available to operators towards the cost of vehicles suitable for use as one-manners on service routes, and Bedford built the Y-series, inspired it seems by the conversions done by Yeates. *White Bus* ordered such a 45-seater to have a coach shell and seats, but with a front door operated by the driver. This was the first new vehicle since the Dennis 'Falcon' of August 1939, and it arrived in October 1972 as a YRQ-type with Bedford 4.66 cu. ins. underfloor-mounted engine and a Duple 'Viscount' body (LJB 403L).

So, at the end of 1972 the active fleet consisted of the 1957 Duple-bodied Bedford SBG (NRD 371), the 1963 Yeates-bodied SB5-type (6881 R) and the new YRQ with Duple body. As a result of those incoming, the Ford 570E (YFH 53) was sold in January 1973.

After that the fleet would be all-Bedford for 30 years, other than the one 21-seater Iveco bought in 1997. Commenting some years later, Doug Jeatt noted the reliability and ease of servicing such vehicles, the chassis developments using many common parts. 19 Y-series vehicles would now follow up to 2001, both as bus and coach-bodied examples.

Top - This second view of the red-and-cream livered SB5 Bedford (6881 R) shows it leaving Windsor along the High Street on a service to Sunninghill. The bodies by Yeates were distinctive in styling and quite unlike the typical output of others at that time. Note the 'WBS' already painted in the offside panel, along with the jack-knife folding door, the position of which also helped increase the seating capacity. However, this vehicle was only retained for 3 years.

Middle – Windsor Central Station was synonymous with the White Bus Service for many years, giving a sheltered waiting place for passengers, though with the wind in the wrong direction it went straight through! The author also started his interest in buses whilst waiting there each afternoon from school as he waited for his Thames Valley Route 2 or 53/53a service to Bracknell, noting the White Bus comings and goings of 1962-4. This shot by Phil Moth has the YRQ-type (LJB 403L) emerging out onto Thames Street.

Bottom – Phil also caught the same vehicle entering the Station, with the walls of the Castle behind. That building is indeed a very impressive structure, even though those of us who are local are just used to it, the attraction to visitors can be understood, whether seen from the road or the riverside, and must be one of the most photographed bits of England. Bedford LJB 403L is unusually dirty, so was probably caught after a spell of wintry weather. Note the sign allowing only buses to exit that direction.

1973-4 – The Third Generation

The owners of *White Bus* have always acknowledged the role played by their hired drivers, fronting as they do the operations, whether they be service journeys, contracts or private hire, and indeed some drivers were instrumental in attracting customers for outings. Such a mix of activities has always required dedicated full-timers, along with those who could cover part-time or just for occasional use. The latter were either working at something else, perhaps lorry drivers or for other operators, whilst some were more often found maintaining the fleet or basically in retirement.

Details available for the first half of 1973 show those active at that time were Cecil and Doug Jeatt, Jim Waterfall, Jim Collins, all working most days, along with Ted Waterman (ex-*East Berks Services),* Mick Fazey, Rod Haylor and Tim Webb, mostly on contracts and private hire, whilst Joe Sutcliffe of *Winkfield Coaches* also features at times.

With Windsor being the Royal Town it is, naturally it hosted various State Visits from time to time, and on 3rd April 1973 the Mexican President and entourage were entertained in the Home Park. Such events did bring in extra passengers, but could also upset the bus timings, especially if the High Street was used for a parade from Victoria Barracks. On that particular day the children of St. Michael's School, Sunninghill took a *White Bus* private hire to see the spectacle, with full instructions issued as to drop off and parking places in view of the heightened security issues of that period.

The Spring of 1973 also saw the start of a daily school contract to St. Edward's School, then still in Dorset Road, Windsor, along with the Charters School run. Local schools also featured more in private hires that year, with a number from the Bracknell area using the firm to reach other schools or the Sport's Centre for events.

The local Royal British Legion Branches also hired at times, with the Crowthorne one taking a trip over to Normandy, though actually the place of that name in Surrey, rather than the D-Day beaches! It was indeed very notable how many hires came from Social Clubs, many of which were actively supported by employers to bring workers together, particularly the large ones in Bracknell, many of whose workforce had relocated with the company. Amongst those noted that year were the MacFisheries Head Office off to Monkey Island (a nightclub near Maidenhead), and regular hires through Priestwood-based Ted Colling, mainly to football fixtures at Wembley Stadium.

On several occasions larger hires were shared with *Winkfield Coaches,* such as 2 coaches for Bullbrook Cricket Club to the London Palladium, whilst on another day the higher capacity Bedford YRQ (LJB 403L) was loaned to the Maulers, complete with driver Ted Waterman, taking the Newtown Singers to Knowl Hill from Harmanswater in Bracknell. As noted previously *Nuetown Coaches* helped out at times, which worked both ways when Barry required more capacity.

The coaching aspects of the Duple 'Viscount' body on the Bedford YRQ (LJB 403L) are more appreciable in this rear view by Phil Moth in the Windsor High Street, the rear sign-writing being rather minimal.

Once Doug had settled into his new role he set about exploring ways of making the business pay, one being money available for fuel-duty relief on rural bus routes. Records of the service mileage were sent to Berkshire CC, which brought in a valuable £1146! Over the following years such payments were further enhanced by co-operation with the County to ensure that core public needs could be met, the Council also having a vested interest in keeping scholars on service routes in order not to have to pay for more contracts.

The timings and loadings on the service were studied in detail, with such surveys and calculations still to be found in the archives. Some minor timing changes also became necessary to combat traffic congestion at peak times, as that situation came to affect the area in the '70's onwards. From July 1973 the last journey from Windsor left at 18.15 on Mondays to Fridays, though a 22.00 still operated on Saturdays through to Sunninghill. Some journeys omitted The Crispin and the Woodside loop, but all ran via The Village. On Tuesdays and Fridays buses also extended to Royal Lodge, whilst the 08.05 from Sunninghill still picked up through the Park for the benefit of schoolchildren.

From October 1973 the route was varied to include Rangers Gate and The Village Crossroads as required, whilst from December that year a 19.00 departure ex-Windsor was added, but it ran the straight route via Cheapside, only going via The Village on Saturdays. However, the 18.15 from Windsor would now only call at The Village and Woodside by request on its way through to Sunninghill. The 22.00 Saturday run would now call at Blacknest on request, diverting via Mill Lane after Cheapside, whilst daytime services in the week saw the 13.00 ex-Sunninghill amended to 30 minutes later in compensation for lost *Alder Valley* workings locally. The latter had been making drastic cuts locally, the true folly of the merger of *Aldershot & District* and *Thames Valley* now becoming reality!

Fortunately for the *White Bus* passengers, great effort was put into finding ways to create economies without cutting essential links, an often difficult conundrum. It also soon became clear to Doug that his return might not be as temporary as he had first envisaged, as his Father's health deteriorated. During November 1974 the Company Name was changed to *C.E. Jeatt & Sons Ltd.* in what amounted to Cecil's last official act, as the desired *White Bus Services Ltd.* could not be allocated as it was already in use elsewhere. In reality only Doug would have any serious involvement in the business, though Gerry and Matt did lend a hand later.

Doug also set about improving the income from hires and contracts, though in respect of the latter is should be noted that in 1974 the Charters School run was £9 per day, such have been the ravages of inflation over the years that it would rise annually.

On the private hire front there can be little doubt that Mrs. Jeatt's contacts through local schools and the Catholic community would greatly assist the process of expanding such work, whilst the principal of giving a reliable service at a sensible price would result in return business. St. Edward's School made a number of bookings that year, including one combined with a river cruise on Salter's Steamers, whilst the choirboys were taken to a service in Portsmouth that July. From Lambrook School, Winkfield Row came a request to ferry 130 children to carol concert practice at Winkfield St. Mary's Church, then on the day of the event with parents present on the Sunday 170 were carried there, but only 90 back to the school. Such jobs would lead some years later to *White Bus* being selected for a daily contract when the school merged with Haileybury of Windsor at Winkfield Row.

As Doug had been absent from his sea-going duties for some time, it became necessary for him to return for a 6-week stint in order to maintain his 'ticket', as many forms of permit-to-work were then referred to. His cousin Terry Dwan had gained his PSV Driver's License, and together with Richard Ludlam, who we heard of back in 1968, helped to run the firm whilst Doug was away. From that association came several strands of future history, as Richard would duly marry Doug's sister Kate, though then with a TV Sales and Repair Shop in Silwood Road, Sunninghill, whilst Terry found his license of use when working as a 'Special' with the Met Police for moving impounded vehicles of that size. Apparently though, the pair got a bit carried away and ordered a bus in Doug's absence!

The new vehicle was once again a Bedford YRQ-type, but carried a Willowbrook 'Expressway' 45-seater body, a design with a coach shell but bus seating and an inward-opening front door, a type developed to take advantage of the Bus Grant. Its arrival certainly gave Doug some sleepless nights, though the grant-aid helped soften the blow and, after 24 years in use at least it proved its worth!

The 'surprise' YRQ-type (SNK 255N) is seen passing through Cheapside en route for Windsor in this photo by Tony Wright. This vehicle was also a popular one for excursions and well liked by the drivers.

The health of Cecil Jeatt gave out on 10th October 1974, when he passed away at Heatherwood Hospital, which in more recent times also marks the terminus of the bus service from Windsor. The next letter to the Traffic Commissioner informs him of the change, with Doug signing as Manager. By coincidence that year also saw the passing of Maurice Nugent, one time friend of Cecil and *White Bus* driver/conductor.

1975-7 – Fares & Experiments

In this period no additional vehicles were purchased, the emphasis still being on making the operations as economical as practical, costs in the industry as a whole being adversely affected at that time. Better scheduling of workings, which saw vehicles taking in service and contract runs throughout the day, led to the release of the diesel-engined Bedford SBG (NRD 371), which went in July 1975. That was followed by the Yeates-bodied Bedford SB5 (6881 R) in October of that year. Later that year the withdrawn AEC 'Reliance' (VXP 508) was finally disposed of, a combination of its poor braking and body issues defeating plans to resurrect it. Indeed, Doug recalls how surprised he was that when the 'scrappies' turned up to collect it they started it up, with clouds of smoke, then drove it away just as it had been laid up all those years!

Not long after Cecil's passing the old Bell Punch tickets gave way to Setright Ticket Machines, which issued roll-fed tickets carrying the date, fare stage, class of fare and value, all set by rotating dials, at first wound by hand but later modified to powered use. One of the White Bus machines is seen with the common modification to hold the roll-cover on.

By January 1975 the level of fares was the subject of meetings with Berkshire CC, as they had lagged so far behind those of *Alder Valley* and *London Country* by some 50%. Doug concluded that he could not impose such an increase on his passengers in one go, so it was agreed that fares would rise by phases over the year. The County did however accept a higher increase for scholar's season tickets, demonstrating the importance of that facility to both parties. It is also worth noting that relations with *Alder Valley* were still most civil, the Traffic Manager then being Mr. T.A. Harrison.

For 1975 we find the daily driving duties covered by Jim Waterfall, Jim Collins aided by Doug Jeatt, as well as part-timers Reg Brooks, Mick Fazey and Peter Horton. The last named was also a Concorde pilot with British Airways, which had a number of crews on short time due to financial difficulties. He already had a PSV License, having married the daughter of Ben Shreeve (*Belle Coaches*) of Lowestoft in Suffolk, flying The Queen to Canada and The Middle East, as well as holding 5 World Air Speed Records.

Doug continued to explore ways of getting support from Local Authorities for the service, whilst also bringing more methodical ways to the office, setting up proper Terms & Conditions of Employment, as well as a pension for full-timers. Industry training and health and safety grants were also taken advantage of, but at that point of time Doug was still undertaking the bulk of the vehicle maintenance himself.

There is also a sense from the diaries of the time that from 1975 the daily relationship between *White Bus* and its neighbouring *Winkfield Coaches* became more fluid, with the latter providing a vehicle to cover the bus service or other commitments when required, and it was not unknown for Doug to drive on a *Winkfield* job. At other times the capacity called for a pooling of resources, one example from January 1975 being the mass outing of Meadowvale Junior School, Bracknell to see Peter Pan at the New Theatre in Oxford, which required one vehicle each from *Winkfield Coaches* and *White Bus,* with a pair from *Nuetown Coaches.*

The Summer of 1975 saw Charters School going to the swimming baths in Broad Lane, Bracknell, still an open-air pool, whilst an attempt to address the needs of teenagers in the local villages saw an evening job each week calling at various points from Ascot into Windsor Youth Club in Alma Road, paid for by the Parish Council at £7.50 for the 19-mile return trip.

There were many school-related private hires, again seeing both firms often out on the same day, London museums and theatres still most prominent, but also some newcomers such as the Weald & Downland Open Air Museum at Singleton in West Sussex. Other outings saw parties going to see Billy Smart's Circus at London's Olympia, whilst another notable event was a hire to *Bee-Line* (as part of the ill-starred *Alder Valley* had become), taking a coach from Windsor to the Mount Pleasant Hotel near Kings Cross in May. Another job would become a regular annual event, when the Royal British Legion Branches went to the Royal Military Academy at Sandhurst, the local streets filled with 100's of coaches from all over.

A surprising number of the hires came from schools in Bracknell, as the town still lacked any real capacity, and over at Wokingham the long-established operator *Brimblecombe Bros.* disposed of its last large-seating vehicle in March 1976 in favour of minibus work. As it happens there was also a contraction in the general coach scene over at Reading, whilst the *Windsorian* star was waning, even though the name continued, all of which gave new opportunities for *White Bus* in areas unlikely only a decade earlier.

On the services side much thought continued to be extended to fares, but there was the usual issue of to what extent such increases might put passengers off the buses, especially as more now had car ownership in their grasp. Indeed, although Michael Clancy used the bus after leaving his original Bracknell–based job in 1975 to a new one in Windsor, he also passed his driving test in 1985, so was another lost passenger.

No changes occurred to the fleet during 1975, but the recurrent faults with Bedford YRQ-type (LJB 403L), culminating in a wiring-loom fire out on service in October 1976 led to its relatively short 7-year stay.

This photo taken by Tony Wright in December 1974 shows just how close-by the two Winkfield-based firms were. A Plaxton-bodied Bedford VAL70 (PLG 702H) of Winkfield Coaches stands beside the Bedford YRQ of White Bus (LJB 403L), with the Yeates-bodied SB5 (6881 R) to its right and YRQ-type (SNK 255N) to its rear. Indeed, the author believed for many years that Winkfield Coaches was just a name used for the coach activities of White Bus!

For 1976 we find driving duties shared by Doug Jeatt, Jim Waterfall, Jim Collins, Mick Fazey, and Peter Horton, with Richard Ludlam included from October.

Much of the private hire that year still consisted of school-related trips, though with many more from the Bracknell area, plus regular bookings from the Royal British Legion, Lawn Tennis Club, Brownie Packs and the College all based in that town. A longer run saw a party from Charters School taken to the Isle of Man via the Merseyside ferry terminal. Staff outings still continued, with Oakleigh Animal Products, and also the Safari Park staff, whilst there was the Garth Play Group, the first reference to that type of work which would certainly increase as many more such organisations came into being.

There were numerous instances when a vehicle was loaned to *Winkfield Coaches* or *Nuetown Coaches,* on Sundays in particular, whilst vehicles or drivers were exchanged on a frequent basis. *White Bus* covered the Rathdown Industries contract for *Winkfield,* taking workers to their light engineering works on London Road in Ascot, whilst taking staff to Safari Park from Slough was covered for *Windsorian Coaches.*

Another new operator in the Maidenhead area, *Barry Reed,* also requested cover on his Maidenhead School runs to Taplow Sports Centre when the capacity was too many for his smaller vehicle. Also, on Monday 6th September *White Bus* obliged with an impromptu rail-replacement service between Windsor Riverside and Staines Stations after a train breakdown.

As noted the siblings of Doug Jeatt had not been involved in the business and, indeed, such activity was of a limited duration. In February 1977 his brother Matt passed his PSV Driver's badge, whilst in April of that year we see Richard Ludlam once again listed amongst the drivers, as was newcomer Tony Smith. Matt's involvement came after Doug married Christine Janet ('Jan') Chapman, and in 1978 he went back to sea by way of the couple having a world cruise on PanOcean's chemical tanker 'Post Challenger' for 6 months before starting a family, which duly comprised of Thomas William Rule (1979), Claire (1980) and William (1984).

Towards the end of May the passenger loadings on later journeys were examined, particularly now that the 'television culture' was reducing attendances at cinemas, theatres and of course use of the buses. The 19.00 from Sunninghill on weekdays, along with the Saturdays-only 22.00 from Windsor were particularly affected, so an application was made to delete those journeys. The Traffic Commissioner asked for figures, which between 2nd April and 11th June showed that each trip only averaged 7 users, both ending in July. As it was the 22.00 on Saturdays was actually worked by the same bus ex-Sunninghill at 19.00, so a whole return duty was saved, including a wasteful lay-over.

Through many years the camera of Tony Wright has captured the White Bus fleet at work over the very varied routes, especially within the confines of that unique area the Great Park. Careful study of the timings, special permission from the Crown Estates and some co-operation from the drivers has produced a catalogue of interesting photos, and here we see Bedford YRQ-type (SNK 255N) in typical Great Park landscape.

Meetings were held with the County Council in order to explore ways of making the service pay better and, it has to be said, ensure that it could continue. As the result of such discussions Doug contacted the Traffic Commissioner in late May 1977 to explain some experiments he wished to try out with that in mind. He considered that making the service better known as a potential tourist attraction might help, so a Round Trip ticket was proposed, which would run experimentally through to the end of September. It was have a fare of 60p return on certain off-peak journeys, but would not be available for the normal return passenger who took time before coming back. Instead, such users would have to return by the next departure. Doug explained that the exceptional beauty of the area travelled through would be the attraction, and a chance to sit back and enjoy splendid and changing vistas. He also noted that such support could help make the service a viable one.

The journeys eligible for such returns were stated as –
Tuesday/Friday/Saturday from Windsor 9.15 and 11, with returns from Sunninghill at 9.35 and 11.35
Daily from Windsor at 12.15 and at Sunninghill 13.30
Saturday only Windsor at 16.15 and Sunninghill 16.50

At the same time, and intended to encourage extra riders for the school Summer holiday period, a special reduction in fares for children aged between 5 and 14 years, who were accompanied by a fare-paying adult, meant they could travel for half-fare instead of the usual calculation then based on 75%, both of these measures being approved by the Commissioner.

A variety of covers were again provided throughout 1977, with *Barry Reed's* contract to Charters School, whilst on occasions Bedford SB8 coach (81 DBL) is found on the *White Bus* contract, having been swapped for the day with *Winkfield Coaches* for YRQ-type (LJB 403L) for a private hire job, whilst *Nuetown Coaches* borrowed 'SNK' for one day.

Swimming runs still featured, such work fitting in nicely with other contracts, and for 1977 included Windsor High School (as the former Girl's Grammar was now known) and Royal Free Junior, both to the Windsor Indoor Pool down by the riverside. Charters also had an exceptional outing to the Ford Motor Works at Dagenham, and indeed the theme of future employment for pupils often drove the purpose of such visits, especially to more local employers.

It seems word had spread from the Brownies to the Cubs locally, resulting in more packs hiring for visits to pantomimes, whilst social clubs booked for the London theatres or a growing trend of attending live shows at the various TV studies in West London. Jobs for local Play Groups also grew, with the Adventure Playground now opened at Easthampstead, whilst a new popular venue for younger audiences was the Disney Odeon, a former cinema in St. Martins Lane in London now showing exclusively Disney productions. Another cinema run saw Lambrook School returning with business, the job requiring 2 vehicles so 'with Dick' on that occasions, plus another joint venture to the Rushmoor Arena at Aldershot.

One of the longest days yet covered saw 'SNK' leave the Northumberland Avenue Social Club in Whitley Wood at 8am down to Barry Island, not returning until 11pm, a long day for Mick Fazey! Again, it is strange that *White Bus* was preferred over the Reading-based options, but that job was repeated annually for several years. Staff outings in 1977 included ICL Beaumont to South Ealing, ICI Jealotts Hill to the theatre in Drury Lane and Crellon Electronics of Slough visiting the associated Marconi's Social Club at Farnborough.

For schools the London Museums were still favoured, but there was a trend to more site visits, such as the Priestwood School tour of Caesar's Camp, Silchester Roman Town, ending up at Reading Museum to study the Roman finds associated with those excavations.

The covers for other operators make interesting reading, as they often contain routes that would later be covered as contracts by *White Bus*, whilst we have already seen that contact with schools never did any harm when operators were sought for further outings, whilst trips for Charters School had steadily expanded with the capacity of that school over the years.

1978-81 – Steady As She Goes

January 1978 saw further considerations on fares, and the extensive calculations survive in the archive to show how much analysis went into such matters. The clawing-back of the slippage of fares compared with the neighbouring concerns still needed action, but on the other hand ways to reward regular users would hopefully halt the loss of existing patrons. In respect of the latter new 1 or 3-month Adult Season Tickets were introduced from early 1978, based on a rate of twice the single fare less 10% per day. No fleet changes took place during 1978, the small fleet being kept busy throughout the day by switching vehicles from service, contracts and in-fill hires to use them more fully and avoid dead mileage.

On the service route from 17th July 1978 the 08.00 ex-Sunninghill was noted as leaving sharp, formerly at 08.05, the time advanced in order to take in The Village but still make Windsor in time for the schools. However, by 1978 it is noticeable that the Charters School run previously subbed-out in the mornings to *Imperial* was now only operated by that firm, though in due course it would come back to *White Bus*.

Covers between the firm and *Winkfield Coaches* were again commonplace in 1978, with the latter's run for Rathdown Industries covered for a whole week, whilst some small-capacity jobs were covered by *Barry Reed* rather than turn them away. The swimming contract expanded to take 3 days per week, with Holy Trinity (Sunningdale), Braywood (Oakley Green), Datchet St. Mary's, St. Edward's RC (Windsor) and Kings Court (Old Windsor) all now added.

January 1979 was badly affected by snow during the third week, aggravated by a strike at that time by the Public Sector worker's, so the roads did not get gritted and both Cheapside and Sunninghill were impassable for a few days, the bus service being reduced to run between Windsor and The Village only.

As commented on a little earlier, the Duple 'Viscount' bodied Bedford YRQ (LJB 403L) had given quite a bit of trouble, so on 29th August a new replacement arrived. It was a YLQ-type chassis (HRO 958V), but this time carried a Duple 'Dominant' service bus body with front entrance and 45 seats. The outgoing YRQ then became a car-transporter, with ramp through the rear end, a fate often associated with the lower floor 3-axle Bedford VAL's in the style of the 'Italian Job'!

The incoming Bedford YLQ-type (HRO 958V) also carried the yellow ochre band, and is seen on its way to Woodside soon after delivery. It would remain with White Bus for 24 years, so was a good investment.

Another new part-timer joined the drivers from 1979, being Mike George, whose main job was as a Boeing 747 pilot for British Airways, and he worked for the firm for some 3 years.

The Charters School contract was now £18.50, which is double that for 1974, which illustrates how inflation was affecting costs throughout that decade. And so the task of adding further work continued, with regular jobs now being done to and from Cumberland Lodge for the Law Society courses held there, more Youth Club work for Old Windsor Methodist YC, and for Orchard Lea School, founded as a Domestic Science and Girl's Finishing School by Constance Spry, just around the corner from North Street on the Drift Road at Winkfield Place.

From January 1980 an off-peak lower return fare was introduced on services after 9am and before 4pm on Mondays to Fridays, a facility also then being used by the local larger concerns. February saw another round of general fares increases, mainly caused by inflation, with 15% required at that point, whilst the Charters contract now reached £21 per day.

In January 1980 Vivien Mauler passed away at the age of 69, though her husband Dick continued with the *Winkfield Coaches* operations for a further decade.

That year saw further new ground broken as the use of vehicles free from daily school-related work was taken up by Summer holiday cover for Heathfield Girl's School, Ascot and for *Tent-Tours,* an operator based in Windsor and Maidenhead, taking foreign students to London etc. Covers for contracts now extended to Bracknell (Bus Station) to ICI at Jealotts Hill for *Hodge's Coaches* of Sandhurst, and Abbey School to Maidenhead on behalf of *Tent-Tours.*

Bedford YLQ-type (HRO 958V) is seen a bit later after losing the yellow band, outside Sunninghill School at the terminus of the route in this photo by Tony Wright. The clean lines of the Duple 'Dominant' body made for a bus that was easy to maintain, whilst the Bedford engineering was well developed over the years, so downtime was minimal. Note that the blind display still lacks any service number at that point in time.

The Transport Act 1980 partly came into effect from October of that year, which saw control of fares taken away from the Traffic Commissioners, whilst those objecting to proposed new services or amendments now had to make good their case, otherwise changes could go ahead. Locally the effects were not felt much but in some areas it led to wasteful competition.

From figures collected on usage of the service, it was clear how important it remained to those living in the Great Park, and in the days just before Christmas 1980 there were extra journeys reaching Royal Lodge for the benefit of shoppers. Such journeys proved popular and were repeated in succeeding years.

July 1981 saw another crowd-pulling event in the Park, though this time with Royal blessing, the 'Great Picnic-in-the-Park', held on the Review Ground on 4th/5th of that month in aid of Cancer Research. As the site was just off the main road south of Queen Anne's gate, *White Bus* ran extra shuttle buses between there and Windsor Central Station between 10am and 10pm on an on-demand basis.

That same month saw Doug travelling up to Barnsley to view vehicles on offer from the dealer Paul Sykes, after which Ron Richmond was sent up to Rotherham to collect a used Bedford YRQ-type bus (STL 725J). That had 43 coach seats within a bus shell, termed as 'dual-purpose', with a Willowbrook 001-series body new in 1971 to *Simmons* of Great Gonerby, hence the Lincolnshire registration. Despite already being 10 years old it proved to be a useful addition, staying with the new owner for a further 17 years!

During that year Doug went over to the Halimote Road Garage in Aldershot to collect some former *Alder Valley* stop flags, which were presumably used for the extended route in due course, though no stops were ever erected within the Great Park.

On Wednesday 29th July 1981 the wedding of Prince Charles and Lady Diana Spencer took place at St. Paul's Cathedral, and a National Holiday resulted in no bus service operating that day, compensated for by the vehicles taking members of the Royal Household from Windsor to London. Later that year Dick Mauler married Janet Taylor in a rather more low-key setting.

Another aspect of off-peak operation explored was that of planned rail-replacement services, which called for Saturday and Sunday coverage, ideal for those non-school days, and in October 1981 Doug went to Slough Station to meet the local BR Manager.

Like the previous trio of vehicles, the incoming YRQ-type (STL 725J) featured a yellow ochre band for the first few years. It was caught by Tony Wright parked on the double yellow lines opposite the entrance to Windsor Central Station in 1984 ready to depart on an Ascot journey via Fernhill.

For the Autumn Term of 1981 the Charters contract had risen to £25 per day, and it is noticeable that Doug covered the run most days between times in the office.

76

Top – *This view was taken on a day when Geoff Lovejoy and Chris Barber were out and about with HRO 958V, and the bus is seen passing the White Horse at Winkfield Row, one of a number of pubs since converted to a food-only venue and re-named.*

Middle – *Further east that bus is again seen outside St. Mary's Church opposite the White Hart, Winkfield, still a pub and frequented by off-duty White Bus staff, and recommended by the author, whose brother runs it. The purpose of the tour is not known, but may have been route training on contract runs, with these locations being short-cuts to school pick-up points. Since then several White Bus vehicles have used the same spot in connection with wedding bookings.*

Bottom – *The third shot of this theme is one of YRQ-type STL 725J passing The Berystede Hotel, once part of the old service route through to Windlesham and Bagshot. This was on a day when the bus had attended a rally with the White Bus Services Enthusiast's Group, done with the blessing of Doug Jeatt, and a chance to hand out service timetables and a potted history of the Company. The Group comprised some employees and some local enthusiasts and attended many rallies, mainly in the South, but also as far as North Weald, Duxford and Warminster.*

There was still good fluidity with the Winkfield neighbours throughout 1981, with the latter's driver Les Spong loaned to cover the *White Bus* Charters run, whilst they covered an excursion by the Windsor Sea Cadets to London. Indeed, it is no wonder that until this research began in earnest the author was under the misunderstanding that *Winkfield Coaches* was just the coaching arm of *White Bus!*

There were plenty of covers for *Windsorian* and *Tent-Tours,* the latter including its Abbey School contract, which led to some private hire jobs for that Reading school. Other jobs were undertaken for *Alf's Coaches* of Burnham, including Claycots School of Slough trip to Bognor and to Berkshire Agricultural College over at Burchetts Green for Slough-based Castleview School were both covered. Work for *MD Coaches* in respect of the Cox Green contract and Maidenhead FC are noteworthy also, as all the above mentioned clubs and schools would in due course choose *White Bus.*

Other private hires in 1981 included the Crown Estate Office, the York Club, and over at Maidenhead the King George V Club, whilst NEM Insurance, based in Lord Robert's old residence at 'Englemere' on Kings Ride in Ascot took its Christmas Lunch at the Silver Skillett, whereas Sperry's SC chose a skittles night at the Dog & Partridge at Riseley, south of Reading, Heavy snow on Saturday 12th December meant the bus service could not reach Sunninghill, but Doug went in his car 'XBW' and took 3 people to Windsor!

1982-5 – One Returns To Ascot

No vehicles would enter or leave the fleet during 1982-4, all the more remarkable as more business was undertaken. 1982 was also the first year that the firm placed an advert in 'Yellow Pages', then the foremost form of business advertising, *as shown below.*

An interesting application was made during February, when *Windsorian* sought permission to run a seasonal service through the Great Park from Windsor (Castle Hill) to The Savill Garden. That operator had been re-structured after the sale out of the original family, and the intention was to operate 3 times per day in respect of Saturdays/Sundays/Bank Holidays between 3rd to 25th April and from 2nd to 31st October, but with a daily service for 1st May to 26th September. A return fare of £1.50 for adults, £1.20 for OAP's and £1 per child would be charged, along with the relevant price of admission to the gardens, effectively making it a localised excursion, the proposed times being –

d. Windsor	a. Garden	d. Garden	a. Windsor
11.30	12.00	12.05	12.35
14.00	14.30	14.45	15.15
15.45	16.15	16.30	17.00

White Bus responded to the proposal with several observations, which if met would result in no formal objection to the planned operation, these being that the price for the trip must be inclusive of admission to the gardens, that no intermediate stops be permitted, and that timings should avoid clashing with scheduled buses on narrow sections of road in the Park. It started with open-toppers as used on the Windsor sightseeing, but low trees were an issue, after which covered buses were tried, but it duly ended that September.

On the staff front, Jim Collins went full-time from June 1982, and Mick Fazey also during the following March, whilst there are some mentions of Jack Edwards, who also drove for Dick Mauler at times, along with a few turns by Barry Devaney.

The Charters contract had risen to £30 per day by September 1982, when a new daily run from the Park to serve St. Peter's School, Old Windsor and St. Edward's School in Windsor started at £18 per day. The swimming had further grown to now cover St. Mary's, Claycots (Slough), Holy Family, Parlaunt Park and Marish (Langley), and Clewer Hill School, the scheduling allowing one vehicle to handle several runs in the same area for the hour-long sessions. School outings continued to build on a mixture of both state and private establishments, with Ascot-area Swinley, Ascot Heath, Papplewick and Heathfield, Bracknell-based Brackenhale, Garth Hill and The Pines, along with Holy Trinity, Sunningdale and Cheapside, St. Edward's and Oakfield (Windsor), Eton College, Langley Grammar, plus Newlands, St. Piran's and Oldfield (Maidenhead) and the Windsor-based Brigidine Convent and Oakfield School all added to the list served. A particularly exceptional task took 'SNK away for 5 days on behalf of St. Peter's, Burnham, to Swanage, with outings to Church Knowle, Portland Bill, Corfe Castle, Bovingdon Tank Museum, Poole Harbour and Studland Beach, the first of many such extended tours to be covered.

The significance of this photo will become more apparent further down this section, but Tony Wright saw Bedford YRQ-type (HRO 958V) on one of the bus service journeys serving Blacknest Gate. On such trips the bus diverted from Cheapside along Mill Lane in order to reach that gate, after which it ran by way of London Road A329 to pick up the route again at The Cannon crossroads. Such encounters were not by chance, but by careful study of the time-table.

Another 5-day school trip saw Maiden Erleigh at Earley going to Westward Ho! In North Devon from 23rd May, whilst non-school jobs were for Reading University, BMW Bracknell visitors via Heathrow, St. Paul's United Reform Church, Silwood Park (Imperial College), Cranbourne Cubs, Bracknell Athletic Club, The Rise Youth Club and Woolworth's, Wokingham. A special shuttle bus service was provided between Chobham Road car park and Sunningdale Golf Club on Friday 3rd September from 11 to 2 and 4.30 to 7.30.

Alder Valley had been steadily reducing its local bus services in the Ascot/Sunningdale/Sunninghill area, but from September 1982 it stopped running down the High Street at Ascot on the 'mainline' route between Reading and Windsor, turning instead at Heatherwood roundabout.

That led to meetings between the County Council and Doug Jeatt, resulting in a return to Ascot for *White Bus* after an absence of 46 years! The revised route would essentially be the same through from Windsor to Sunninghill, after which it ran as Sunningdale

(School) – Sunninghill (School) – South Ascot – Ascot (Station) – Ascot (Horse & Groom – still referred to as such despite the name having changed!) – Ascot (Heatherwood Hospital). Although buses now ran again along the High Street it was the opposite way to which *White Bus* had been when the service was sold to *Thames Valley* in 1936, though actually as it had been <u>originally</u> under Ackroyd.

In order to accommodate the extension the first bus now started from Fernhill at 7.30am and went via Ascot Gate – Cheapside – The Cannon – Sunninghill (Crossroads) – Ascot (Horse & Groom) – Ascot (Heatherwood Hospital), where it formed the 7.46 over the full route through to Windsor to arrive at the traditional time of 8.35am.

Some journeys went more directly from Windsor along the main road along the edge of the Great Park, past Fernhill, then as per the 7.30am departure, whilst several also took in the Woodside loop from The Crispin and then straight on past Shepherd White's Corner, past the western edge of the racecourse direct to Heatherwood. Also on the daily service Blacknest was only scheduled to be served twice, but certain other journeys could be requested to call there, with a similar arrangement for those wishing to go to or from Royal Lodge.

In order to maintain the schoolday links, the 15.30 ex-Windsor ran only on school dates to cover The Village – Royal Lodge – Ascot Gate – Cheapside, whilst the daily 16.15 (also used by some children)

ran from Windsor as Queen Anne's Gate – The Village – *optional* Royal Lodge – Fernhill – *optional* The Crispin – *optional* Woodside, *optional* Shepherd White's Corner – Ascot Gate – Cheapside – *optional* Blacknest Gate – The Cannon – Sunningdale (School) – Sunninghill (School) – South Ascot – Ascot Station – Ascot (Horse & Groom) – Heatherwood Hospital, the timetable noting that buses may run up to 10 minutes later if taking in optional diversions. The same optional points also applied to the 17.05 and 17.50 from Windsor (not Royal Lodge on 17.05), so when the service commenced on Monday 20th September 1982 journeys varied between 24 minutes via the straightest route and 45 for the full works!

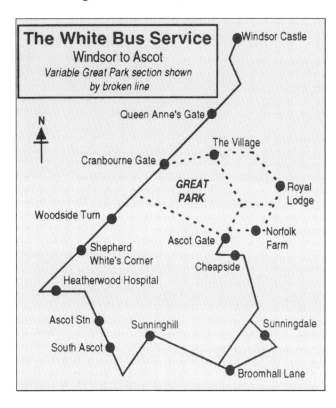

The last departure from Heatherwood Hospital left at 17.10 and arrived at Windsor 17.42, forming the 17.50 right through, which then ran dead at 18.22 back to the North Street Garage, the timetable being similar on Saturdays, but lacking an afternoon trip and that operated on schooldays. Apparently, there was a rule that passengers could not be carried only between the Hospital and Sunningdale (School), due to the *Alder Valley* Route 197, but how well that was observed or appreciated locally is not recorded.

From 1983 there was a notable increase in the use of outside facilities to supplement to capabilities of the North Street premises, with 'STL' going to *Alder Valley* at Maidenhead and 'HRO' to John Lewis at Southern Industrial Area in Bracknell for the fitting of a Tachometer cable, whilst *Moore's* over at Clewer Hill continued to do steam-cleaning. Whilst on the subject of fleet maintenance, the sign-writing was for many years done by Les Clarke, though after his sight deteriorated, Chris Bevan of Bevan Signs took over and continued, later using vinyl lettering.

We have already seen some photos by Tony Wright. He first became interested in buses just before WW2, and here he gives us an insight into how he came to take photographs, those of White Bus making this history so more complete and enjoyable to compile.

Tony's first bus influences were *Aldershot & District* in the Hindhead area, particularly his favourite Dennis 'Aces'. The first camera came in 1946, a Rajah 6 funded by his Mother's tea coupons, but with only 8 exposures on 120 film, his first bus photography.

Leaving school in 1948 he worked in an ironmonger's at Churt until National Service in the RASC as a tactical sketcher from January 1951 to 1953, after that being employed as a cartographic draughtsman by the War Office at Park Royal in North London, then to the MOD at Feltham through to retirement in 1992.

As both these locations had much of railway interest, many photos of the end of steam on British Railways during the 1960's were taken, often appearing in magazine articles or books, by then using an Ensign Selfix 820 and taken as far apart as Penzance and Inverness. From the early 1950's he did his own black-and-white processing, along with some professional photography for weddings, the colour work on Rolleicord 120 film camera.

He married Jeanette in 1961 and they bought a house on the old Langley Airfield east of Slough, which was new territory to them. However, Tony was pleased to find that Windsor offered some interesting independent operators, being particularly fond of the Moore's *Imperial Bus Service*, following operations through to the final evening in January 1987.

In June 1983 he by chance came across 'HRO' in its white and yellow livery at Heatherwood Hospital, and his attention turned to the activities of *White Bus*, soon discovering that the route through the Great Park offered great scope for superb locations in which to portray the fleet, especially with the deviations of the basic route to lesser known areas of the Park, which are ably demonstrated by the superb selection of shots included within this book. In order to capture such vehicles on the move Tony got a 35mm Pentax MZM, which he still uses for his excellent and well thought out photos right through to the present day operations of *White Bus Services,* never succumbing to digital.

Tony had for some time considered that the Company deserved a book, both he and others already knowing the author's work on *Thames Valley* and *Newbury & District*. So with Doug Jeatt's kind assistance the rest, as they say, is now history, as presented here in print.

All photographs included from this point onwards are by Tony unless stated otherwise in the caption.

Top – For over 80 years the children of the Great Park have relied on the White Bus Service to get to their secondary schools, both in Sunningdale or in Windsor, and a number of them are seen alighting the Ascot-bound afternoon bus by the large open green of The Village in October 1992. Bedford YRQ-type SNK 255N carried the 'Expressway' style of 45-seat body by Willowbrook and was bought new in 1974, remaining in the fleet for 24 years. The group of cottages around the green has a timeless feel, though they were built in 1948.

Middle – Another notable feature of the Great Park are the distinctive gates at various points on the perimeter, most with a lodge of individual design. Here we see SNK 255N negotiating Prince Consort Gate, with its large lodge constructed in 1862. The small hut in green is one of many accompanying the gates, sometimes manned by liveried keepers. The bare trees reflect this photo was taken in April 1995, whilst the width of the gate is quite tight, with buses becoming even larger in due course.

Bottom – A White Bus threads its way through the fields of rape-seed, then a new crop in local fields in May 1998. Bedford YRQ-type HRO 958V is seen on Reading Road between the Copper Horse and Sheet Street Road. This 45-seater had a Duple 'Dominant'-style bus body, and has the grey skirt panel adopted by that date. This was a new purchase in 1979 and would also see 24 years in use, the penultimate bus of that make in the fleet.

Top – *The most low-key of all the gates is Ascot Gate, which is only accompanied by one of the little huts. However, it does have more significance each June as the way the Royal Procession leaves the Park on its route to Royal Ascot Races. Bedford HRO 958V is seen in April 1998 on its journey in the opposite way from the Sunninghill Road into the Great Park, where it leaves the traffic of east Berkshire for the sylvan calm of ancient trees and neat grass verges, soon to reach The Village.*

Middle – *Given the rather scattered nature of workers cottages over the Park, it was provided with a school from 1845, built close to The Village at Mezel Hill, its buildings blending in well with others nearby. On this journey from Ascot we see Bedford YLQ HRO 958V passing the Royal Schools. However, the schools only catered for infant and junior classes, and in earlier times the boys grew vegetables and the girls prepared lunch as part of their education, whilst the schoolmistress was given a cow to provide the children's daily milk!*

Bottom –*This fine vista across the stands of woodland and individual field trees is centred on the Stone Bridge, necessary to pass over a natural spring and built in 1829. It is the single largest structure of its type in the Great Park, and Bedford HRO 958V is seen travelling away from the Copper Horse and towards the Royal Lodge in this view of September 1994. Since 1979 this has been part of the Deer Park, and traffic restricted to authorised vehicles only.*

Top – This October 1993 photo shows HRO 958V setting out from Windsor to Ascot, with the main public entrance to the Castle to its rear. It is entering Peascod Street, then still used for restricted traffic but now fully pedestrianised. To the left is Barclays Bank where a Carry On film was based in the upstairs offices, the town often being used by the nearby studios of Bray and Pinewood for filming. Note the grey skirt now adopted for the fleet, which made sense considering the rural nature of the route.

Middle - On the route from Sunninghill to Ascot Gate lies the largely linear village of Cheapside, one of a number of settlements just outside the Park area. Once again the White Bus provided a valuable link for schoolchildren, work journeys and shopping for the locals, as well as links to the rail stations at Ascot, Sunningdale and Windsor. During WW2 the nearby Sunninghill Park was used for training by the USAF, many of whom frequented the Thatched Tavern, and here we see HRO 958V on its way towards Ascot in this May 1998 view.

Bottom - Bedford YRQ-type STL 725J also carried a Willowbrook body, but had 43 coach seats in a bus shell and was built in 1971, coming to White Bus in 1981. By this September 1984 view it had this livery of white with a yellow band, though it was duly all-white and finally white with the grey skirt by 1998. It is seen waiting for pupils from the Marish Junior School in Langley, which had a regular booking for swimming lessons, work of that type fitting in nicely.

Top – Willowbrook-bodied Bedford STL 725J is seen approaching Norfolk Farm in the depths of the Great Park and out of the public access, such journeys being for schoolchildren at the appropriate times. The baring trees show this as a November view in 1997, whilst here we have examples of work on the oak tree, whilst the mixed stand behind the bus reflects the poor soil here. Note that this track is not up to the general standard of roads in the Park, and can get muddy at times.

Middle – As with a number of other pubs associated with the White Bus routes, The Wells at the junction of the London and Cheapside Roads, just east of Ascot, has now become an ethnic restaurant. In this April 1994 shot we see Bedford STL 725J working an odd journey which took it direct from the Cannon cross-roads to Ascot High Street. The bus stop is for services in that area emanating from Bracknell, which are in continuation with links created many years before by The Thames Valley Traction Co. Ltd.

Bottom – Typifying the helpfulness of White Bus drivers over the decades, here we see passengers laden with extra Christmas shopping leaving the bus in December 1993. The Royal Lodge gate-houses are the backdrop at the junction for Bishops Gate. Journeys still reach this point when required, but at that time were still scheduled daily operations. From here the bus ran past the Royal School, Cumberland Lodge and Dukes Lane before re-emerging onto the public highway via Ascot Gate.

Top – 'STL' is seen again at Ascot Gate, or rather <u>in</u> the gate! The accident took place in April 1998 when, apparently the gate began to close, the wet surface meaning it could not be pulled up in time. Over the years there are only about five such incidents recalled and considering the daily use of so many relatively narrow entrances over the years, not too bad a total. However, on each occasion an invoice followed from the Crown Estate for the necessary repairs!

Middle – JMJ 633V was a Bedford YMT-type which would serve White Bus for 21 years, being new in 1980 to Moore (Imperial) of Windsor and coming to Winkfield 7 years later. It carried an 'Express IV' 53-seater coach body by Plaxton, and here we see the interior view looking rearwards. The vehicle saw use on the service as well as contracts and private hire, the Bedfords being quite straightforward to maintain, and it was joined by other similar vehicles. Compare this interior with other later ones shown on pages 118 and 124.

Bottom – In another view of the pink-painted gate-houses of the Royal Lodge, we see JMJ 633V on its way towards the Copper Horse and then to Windsor. The trees are now in full leaf, and each location in the Great Park takes on a different look with the seasons in an ever-changing palette. The tall trees to the left now echo to the squawks of the parakeets now in the Park, though other native rarities such as Hawfinches and Wrynecks might also be found if you are very lucky.

Top – Taken on the same journey we now see 'JMJ' at the Copper Horse, in fact made of bronze and erected in 1831 in memory of King George 111, which forms the finale of the 3-mile Long Walk from the Castle in Windsor. Perched atop of Snow Hill it also has good views southwards over the Park, and has always been a popular spot to start rambles throughout the area to Virginia Water and Savill Garden, or closer by to Bishops Gate, Smiths Lawn and the other equestrian statue of Albert, The Prince Consort.

Middle – Peascod Street in Windsor had its vehicular traffic removed in stages, and we see 'JMJ' after the area from the castle end down to William Street had been closed. New shelters were erected just out of sight to the left, with the bus entering via Victoria Street, then departing as seen here along William Street. In due course buses would no longer enter this area at all, the new focus for White Bus now being in Charles Street, a short walk from where this photo was taken.

Bottom – Another local bus operator with roots back to the 1920's was A. Moore & Sons (Imperial Bus Service), based at Firs Road, Clewer Hill. It often worked alongside White Bus on larger private hires and sometimes its steam-cleaning ramp was used, but in January 1987 it ceased operations. Two of its Bedford YRQ-types with Willowbrook 001-style 45-seater bus bodies were acquired by White Bus, but GNH 530N was burnt out by vandals at Firs Road in May 1987 whilst in store.

Top – In pleasant contrast to the previous photo, here we see the other saloon acquired from Imperial as GNV 983N enjoying the scent of daffodils as the Spring unfolds. This bus is on the afternoon school run taking in Norfolk Farm, and it now follows the road beneath High Flyers Hill on the way to Johnson's Pond and across the bridge separating that from Virginia Water lake, and on to Blacknest Gate. Note that the carriageway is marked by white-painted posts as a driving aid in an otherwise unlit parkland.

Middle – Mention has been made of the Deer Park set up with The Duke of Edinburgh as Park Ranger, and centred on the Long Walk. Here we see Willowbrook-bodied YRQ-type Bedford GNV 983N passing one of the electronically operated gates, for which drivers carry a zapper on the bus. It has just traversed the Stone Bridge, where the ground slopes away and there are woodland stands where the deer rest, whilst other such fenced-in stands allow ground-nesting birds such as pheasants to breed.

Bottom – In another view with 'GNV' we see it in the North Street garage in May 1995, at which time it had a Union Jack, and is seen with the blind set for route 24 to Charters School. This group accounted for many years of service to White Bus, with (left to right) the senior driver Mick Fazey, driver Alan Moore, freelance mechanic Brian Lovejoy and his son Geoff, who was also a driver. Alan was indeed one of the family of Imperial fame, being a cousin of the sons.

Top – Near the top of Priest Hill, and a short distance from the Thames at Old Windsor, is the Roman Catholic College of St. John Beaumont. Quite a lot of hire work emanated from that establishment, and Bedford YRQ-type bus GNV 983V awaits a party outside the ornate chapel and residential college in December 1995. Students attending Summer School also required transport to and from the site, along with excursions to London and other venues, fitting in well with school holidays.

Middle – Just a short distance away from the college is Bishops Gate, and again we see 'GNV' as it enters the Park from the Englefield Green area. The gatehouse here is of yet another style, and in this case a bungalow. By then the bus was 13 years old, but it would go on for another 11 years in White Bus service. In all, leaving aside the ill-fated 'GNH', the fleet featured 4 YRQ's fitted with Willowbrook bus bodies, and between them they clocked up half-a-century of service with the Company.

Bottom – The interior of 'GNV' is seen from the rear of the saloon, as it was about to leave from Windsor. The driver is Geoff Lovejoy, who helped his father Brian wash buses etc. in his youth, duly obtaining his PSV License with Beeline before joining White Bus finally in 1993. The bus is well loaded on this afternoon journey, but some trips were less full. The layout is typical of underfloor-engine buses of the 1950's -1980's, with front-facing seats and a parcels rack above.

Top – Another view of the journey serving Norfolk Farm, with the buildings seen in the centre of the background. Bedford YNT-type bus C668 WRT had a Duple 'Dominant' body with 3+2 seating to give a very useful capacity of 63. It was new in 1986 and arrived in October 1995 to give another 12 years of service. Here it climbs the rough-surfaced track from the farm, in the opposite direction to that depicted on page 84. Norfolk Farm had been brought back into production during WW2, whilst also note the simple but effective fencing.

Middle – The school runs take the White Buses over many roads far away from the service routes, and C668 WRT is seen passing the Fleur-de-Lis at Lovel Hill on its way to form the 28 route from Charters School in Sunningdale. It was here that the Great Western Railway stabled its bus from 1905 until the mid-1920's, whilst from 1921 William Rule Jeatt had his initial premises locally as cycle and motor repairer. The pub has since been turned into flats.

Bottom – Royal Ascot Race week in June each year may bring some additional passengers, but the traffic problems are an annual headache for White Bus. In the Great Park itself the Royal Procession causes some delays, whilst the local roads are very busy. This view shows 'WRT' on the roundabout where the High Street meets Station Road, with the racecourse buildings behind the pub, the former Horse & Groom where earlier on we saw the pioneering GWR bus standing back in 1905.

Top – This view sets out to recreate one with Dennis Ace JB 9468, when posed by the newly-planted trees of the Long Walk, with the Castle seen in the mist 3 miles away. Taken in May 2006, it shows the avenue over 60 years later from in front of the Copper Horse. The bus itself is 20 years old at that point, and by then was mainly used on school contracts, so this was probably on a journey by the Enthusiast's Group, comprising of some local enthusiasts and staff and was active 1994 to 2009.

Middle – Here we see the same bus from the other angle on the same day. Note that since it had been caught by the camera on the previous page it had gained permanent school transport signs. The high point of Snow Hill affords good views southwards as well, whilst to most people the Copper Horse appears in silhouette, though on closer inspection it shows George III in the guise of a Roman Emperor and bears a green tinge. The statue was thoroughly restored in 1968 after some cracks were discovered.

Bottom – Seen just past the turning for Bishops Gate, with the Royal Lodge gate-houses behind, this view of C668 WRT also reminds us that the White Bus Services ran all year round. The heavy snows of February 1996 are still in evidence and, although the roads within the Park were not subject to the County Council road gritting, the main network was usually passable in all but the most extreme of conditions, the general lack of other traffic also being helpful for the driver as he makes his way.

Top – Bedford 'WRT' is seen again in the Spring as it leaves through Blacknest Gate. The lodge here has turreted chimneys and is painted pink, being built in 1867. Note also the lamp-posts found at most of the gates, offering a point of reference in an un-lit area. The rider to the right also serves to point out that permits are issued for the Great Park, and in many places riding tracks have been provided. There are also Driving Tracks for horse-drawn vehicles, and at one time The Duke of Edinburgh was regularly to be seen at the reins.

Middle – The outbreak of Foot-and-Mouth disease in 2001 caused widespread devastation to cattle over the country, so very strict controls were brought in on movements etc. In the case of the Great Park the access was restricted to only two gates, one of which was Rangers Gate, seen here with Bedford C668 WRT about to pass over the disinfectant straw matting strewn across the road, with a stock of bales stacked by the roadside.

Bottom – Another type of work which fits in nicely with weekday commitments to school runs is the hiring of vehicles for weekend rail replacement services to cover planned engineering work. Despite displaying Route 01 on the blind, here we see 'WRT' in June 1997 at Slough Station on a journey to Windsor in place of the branch line. A line-side fire had broken out, sending these dark clouds billowing up, but at least the situation gives an opportunity to appreciate Brunel's architecture free from modern distractions.

Top – The timetables only reflect one aspect of the work which goes into running a bus and coach fleet, and in this evening shot of North Street garage we see Bedford YMT coach JMJ 633V in the workshop, with YNT-type coach JNM 747Y to the left and YMT-type RLW 778R on the right, the latter now in full White Bus livery after being formerly a Winkfield Coaches vehicle. The photo is taken in January 1998, but the site had been in use for some 68 years by then.

Middle – This view taken during July 2001 shows Plaxton-bodied Bedford YMT-type coach EAJ 327V about to depart from The Village, with a party of smartly-dressed ladies on a private hire. The Great Park had its own branch of the Women's Institute, as well as Scouts and Guides, whilst the sporting teams based on the York Club led to a steady flow of hirings throughout the seasons to places both near and far. 'EAJ' was new in 1979 and bought in 1985, remaining in the fleet until 2004, and had a Supreme IV body.

Bottom – Another regular weekly booking was taking senior citizens from a day centre based in Datchet to various other local places, and here we see Duple-bodied Bedford YMT-type coach RLW 778R leaving the village over the level crossing. Tony had set up on the footbridge to get this shot, with the notable half-timbered buildings in the background. This coach was built in 1977 and came with Winkfield Coaches in 1990 to serve for another 9 years. It is seen in January 1999 just before its final withdrawal.

Top – *The Great Park is studded with ponds of varying sizes and shapes, most being artificially created by damning local springs and streams. Shop Pond makes a pleasant backdrop to this specially posed photo. In former times these ponds were important for fishing, watering livestock, and as decoys for wild-fowling, and even some Royal boating! In more modern times most have been fenced as we see here. Plaxton-bodied Bedford YNT coach B633 DDW is seen in June 1997.*

Middle – *Though referred to as a pond, the rather more lake-sized Johnson's Pond is situated towards the top of the complex forming Virginia Water, being divided from the main body by the road bridge which B633DDW has just crossed over in another June 1997 view returning from Southsea bus rally. Note how coach bodywork has developed by comparison with 'RLW' opposite, the higher floor line providing more room for stowing luggage.*

Bottom – *Another Plaxton-bodied Bedford YRT-type coach was GHD 668N. It had a 'Panorama Elite' MkIII-style body and had come to Winkfield Coaches in 1983, being built in 1975, passing to White Bus in 1990. However after a while laid up it was finally stripped of useful parts and sold for scrap. It is seen here on 29th April 1993 by The Squirrel being towed away by Gough's Warfield Garage, another firm with a long local history which included coach-building, a bus service and coach operation from the 1950's.*

Top – We have already seen that a White Bus often attended bus rallies with the WBSEG, and here we see a group posed in front of Bedford YRQ-type SNK 255N, the photo being taken in April 1994 when the Cobham Rally was at Apps Court Farm, Walton-on-Thames. On the right is Tony Wright, whose photos make such an illustrated history of the later period possible. Geoff Lovejoy was at the wheel that day, and the board in front shows a history of the firm. The Group produced fleet lists and White Bus mugs designed by Gerry Bixley.

Middle – Having already noted the long lives of most Bedford vehicles at White Bus, we now come to an exception. Although it was a YMT-type with a Duple 'Dominant' 53-seat body, ODL 632R of 1977 was only kept for three months from September 1996, hence its appearance here still wearing the pink livery of its former owner. The bus behind is seen fully below, and was not a member of the fleet, though kept at the North Street yard and maintained there.

Bottom – C911 VLB was a former British Airways staff bus with a Wadham Stringer body on a Dennis G10 chassis, seen here in April 1996 and new in June 1986. It replaced a Mercedes minibus stolen from St. Peter's Hall in Hatchet Lane, where the 1st Cranbourne Scouts met, and which was never recovered. Dave Empson and George Lynham were both Scout Leaders who drove it, also working for White Bus after gaining PSV licenses through the Company.

94

Top – This Bedford YRQ-type was a 1975 model with a 45-seater Plaxton 'Panorama Elite' MkIII body, passing to Winkfield Coaches in 1979, and then into White Bus in 1990 and stayed 13 years. KDT 281P was later used for school runs, and is at the Crooked Billet at Honey Hill, south of Wokingham on Route 39 running from California Crossroads via Nine Mile Ride and St. Sebastian's to St. Crispins School on the London Road, Wokingham. This supported service was registered by White Bus.

Middle – At the opposite end of White Bus school operations we see Doug Jeatt returning from the morning run to Cox Green School, and opposite The George at Holyport, the route serving the various scattered communities in that catchment area. SNK 255N was a new purchase in 1974 and carried a Willowbrook 45-seater 'Expressway'-style body, and it is seen in April 1998 just before its final withdrawal. Note the golden cockerel used by Courage on its inn signs.

Bottom – A number of shots show White Buses at Langley, partly because it is where Tony Wright lives. Here he caught Bedford YMT coach JMJ 633V on a hire to Ryvers School in Trelawney Avenue, for their swimming lesson transport, as a pool was out of the reach of most school finances. Such hires had featured with White Bus back to the late 1930's. In a sad reflection of modern thinking, taking such a photo featuring children now might well result in the attention of the Police.

Top – At the Windsor end of Route 13 we see Bedford SNK 255N in Vansittart Road in May 1998. To the left is the entrance to Windsor County Boy's School, whilst the building behind houses Princess Margaret Rose School for girls on a site between the Maidenhead and Oxford Roads. The Mitre pub is on the right, whilst behind the photographer is another, the Vansittart Arms. These streets often doubled as 'London' in the Carry On films or those with Norman Wisdom made there.

Middle – Mention has already been made of the 63 seats in C668 WRT with its 3+2 arrangement from the 4th row backwards. This July 2002 shows the 'sharp end' of school runs, with pupils from Trevelyan School at Windsor. When I was in school transport my advice to drivers was to get on with the kids, put the radio on if there was one, as only a fool starts a war when all of the opposition is sitting behind him! Some drivers had that knack, or maybe they just switched their deaf-aids off?

Bottom – Our final shot of this first colour section has Bedford coach 'EAJ' later in its career in June 2002. Doug Jeatt is again at the wheel on this first day of operation of Route 88 from Windsor through to Fifield. The coach is at the junction of Dedworth Road and Oakley Green Road by what was then the Nag's Head pub, and it catered for children attending St. Edward's, Windsor Boy's, Trevelyan and Windsor Girl's Schools, with stops at Fifield, Oakley Green, Dedworth and Clewer and called nearby each school.

Some changes to the schedules were required at short notice from 17th September 1984, when the County Education Department informed the Company of some revised school times. That resulted in the 15.50 from Windsor running from Cheapside over the full route to Heatherwood Hospital. Whilst in order that the 7.46 morning departure reached Windsor in time, it no longer served Watersplash Lane, going by way of Cheapside – New Mile Road (by the Race Course) – Winkfield Road – Shepherd White's Corner.

Another issue to raise its head during September 1984 was the re-development of Windsor Central Station, which it must be admitted was then looking rather tired, and certainly not a glittering gateway for tourists. Although the area would benefit from the coming investment by the Madame Tussaud's Group, which would install a Royalty & Railways Exhibit and the first phase of shopping and restaurant there, the bus companies would of course see themselves sent to various unconnected points around the town centre. The author also recalls that at that during that period many of the fine bus stations around the country, and in particular the once proud and well-maintained ones of *Green Line* were being sold off to developers, with passengers relegated to windswept bus stops often on the sides of busy roads, as second-class citizens.

We will take a look at how other work fared during the period 1983-5 a little later on, but despite the evident increase in general operations, no changes took place in the fleet until April 1985. In response to the number of private hires a coach-bodied Bedford YMT-type (EAJ 327V) came in, which carried a 53-seater Plaxton 'Supreme' MkIV body and had been new in 1979 to *Begg* of Middlesbrough. Although being 6 years old already, it would remain in the fleet until 2004, a tribute to the care given to the vehicles.

EAJ 327V seen at St. George's School in South Ascot.

One of the regular maintenance tasks in those days of smoking on PSV's was a deep-clean, with surfaces given a hand-clean to remove the sticky build-up of toxins, as well as spraying the moquette to re-freshen.

The period 1983-5 saw a notable increase in private hires, whilst the willingness to cover for other local operators saw repeated requests from *MD Coaches* and *Carter's Coaches*, both of Maidenhead, *Hodge's Coaches* of Sandhurst, *Tent-Tours* and *Windsorian* of Windsor, *Armchair-Smith's* of Reading, along with those from *Winkfield Coaches*. As some of those were school contract runs, there is no doubt that an insight into such routes could help when re-tendering came around in due course, or as operators faded in what was not a good climate generally for such companies. Other runs were for worker's transport, something which would greatly reduce as more people got cars and employers no longer felt obliged to provide links, as at that time the Beecham's factory at Maidenhead brought in workers from as far away as Reading.

As noted in earlier days, the link between *White Bus* and Berkshire schools for swimming transport had started in pre-war days, the original fill-in work, and over the years the outdoor pools had been replaced by indoor facilities, making it now a year-long activity. The Company was now taking a number of pupils from Windsor and Slough area schools to the pools at Windsor, Langley Leisure Centre and Montem Leisure Centre in Slough, whilst some private schools also included Coral Reef in Bracknell at times. Such worked eventually involved many schools in the area, which will be detailed later.

Indeed, the list of schools hiring for general trips grew so much over the '80's that it would be easier to note those not doing so! On the other hand the old pub-based work, such as darts, had virtually dried up, but local company-based social clubs were still active with hires to theatres and seaside outings, one example being the Clifford's Dairy of Bracknell outing each Summer paid for by the firm for staff and their families, which in 1985 *White Bus* covered to Bournemouth, with others for Racals and Ferranti.

Corporate hires also saw parties of foreign visitors being collected from Heathrow, transfers from hotels and staff attending training courses. Other courses at Cumberland Lodge saw students from the London School of Economics brought down, whilst Lincoln's Inn and Imperial College called for vehicles for local transfers and sometimes further afield regularly. At Christmas there were hires to lunch venues or a show, all handy with no school work taking place. The transporting of a widespread range of Cubs, Brownies, Guides and Scouts still took place regularly over the camping season, that class of work also having to cater for all the camping gear when determining the vehicle size!

Another shot of the first 53-seater owned, Bedford YMT (EAJ 327V), seen here with Joe Sutcliffe who would duly transfer from Winkfield Coaches.

Sporting clubs also featured in the hire jobs, with the football clubs of Bracknell and Binfield for matches away, as was so for the meetings of Bracknell Athletic Club, whilst Windsor Rugby Club did a 3-day tour in the Exmouth area. It should also be noted that there were longer school hires to field centres during this era, taking Mick Fazey away for 3 to 5 days to either Swanage or Pembroke, often at half-term breaks.

Family connections with the Catholic community led to various other regular hires for St. John Beaumont RC College on the edge of the Park and overlooking Old Windsor, and for the Catenian Association, a social group for Catholic businessmen, as well as for C.of.E. Canon Treadgold with transfers to Heathrow for parties organised to visit Rome.

Also of a social nature were the bookings from the North Surrey CAMRA Branch to the Beerex event at Farnham Maltings, picking up in Ascot, Sunningdale and Camberley. From Bracknell the Royal Naval Association picked up from the Prince of Wales pub in Priestwood, whilst CANUSPA (the Canadian and US parents and children's group) arranged flights abroad or a 3-day trip to Torquay, and the Ukranian Association of Slough had a number of days out.

In the early 1980's the Crown Estate Office was at Mount Pleasant off Church Road in Bracknell, and it regularly hired *White Bus* to transfer staff or to run parties around the Great Park and The Savill Garden, whilst the Masonic Lodge at Wokingham chose them to visit the Centre at Uxbridge. The first wedding hire noted occurred in 1985 to Winkfield St. Mary's.

An exceptional hire was for Burchetts Green School, some way west of Maidenhead, which was invited over by Cheapside School in June to watch the Royal Procession pass by to Ascot Race Course. Other hires for notable events were for performers and their staff staying at the Old House Hotel by Eton Bridge, with a coach to take them to and fro to evenings at the Wentworth Golf Club.

During the Summer holidays the Ascot English Course hired to collect students from Heathrow and to go out on trips, whilst the Heathfield Girl's School on the London Road near the Royal Foresters, just west of Ascot, ran a Summer School programme that saw students taken into Windsor, Bracknell and London for activities, as well as church on Sundays, whilst the Windsor-based Haslemere Travel used *White Bus* for some advertised excursions in 1985. Similarly, the Runnymede Hotel booked for Heathrow transfers and sightseeing tours to Windsor, and on the schools front there was an increase in bookings by Licensed Victualler's School, though then still based at Slough.

In respect of staff Ron Richmond left in February 1983 but returned again from December 1985, whilst Mick Fazey went full-time from March 1983. Bob Pratley first appears in June 1985, and in October of that year Peter Ives is noted, with Ted Graham from December, though the main service was still mostly in the daily care of the two Jims, Waterfall and Collins. The first use of LPC Coachworks is March 1985, after which Eric Chambers from there regularly did body repairs, and indeed some driving, for many years, though these days he helps restore older types nearby at Colin Billington's workshops in Fifield.

Looking like a toy bus against the mighty oak trees, Bedford STL 725J enters the Park through Ranger's Gate, the Castle being silhouetted on the skyline.

For the Summer of 1985 a descriptive leaflet was run off, with a map of where the bus went in the Great Park, extolling the usefulness of the route to walkers in particular, though a note was added to check times carefully when planning a ramble, also in German, French and Dutch! The leaflets were put out at the local tourist office in Windsor and at rallies, also noting the 55 years of service, with the following text-

THE NICEST WAY TO GET TO WINDSOR

'If you want to get to Windsor, there is no more delightful way of doing it than on a <u>White Bus</u>.'

That's what people have been saying about the White Bus service for the last fifty-five years......and they're still saying it today.

Six times a day in each direction, a White Bus will take you from a multitude of pick-up points in Ascot, South Ascot, Sunninghill, Sunningdale or Cheapside to Windsor's Central Station and back again in comfort, economically, and through some of the finest scenery of any scheduled bus route in the country.

With fares as low as 70p single (£1.20 return) from Cheapside, the bus will take you either through Windsor Great Park or Windsor Forest. What more pleasant way could there be to go shopping, or perhaps visit friends in Windsor?

Well over a million passengers have savoured the delights of the White Bus service since it started running regular services between Ascot and Windsor in 1930.

Still a family-run firm, the White Bus service has a fantastic reputation for friendliness and unbelievable reliability.

Accept with our compliments your free timetable (see overleaf) and try it this week if you can.

Did you know that you can hire a White Bus anytime, anyday, and at very competitive rates. To find out more ring Douglas Jeatt on Winkfield Row 882612.

1986-9 Imperial No Moore

Throughout the mid 1980's Doug Jeatt attended meetings of the East Berks Transport Committee held at Maidenhead Town Hall, where he continued to keep up pressure for Local Authority support for the bus service, which also earned him the respect of the Officers there.

Despite the evident increase in activities, there were no additions to the fleet in 1986, so the quartet of Y-type Bedfords was kept fully occupied, with the 1971 YRQ (STL 725J), 1974 YRQ (SNK 255N), the YLQ (HRO 958V) and YMT coach (EAJ 327V), both of which were new in 1979.

The year 1986 marked the end of the era in a number of ways locally, all of which had an impact on *White Bus* operations in one way or another.

Firstly, in another stage of implementation of the new Transport Act, the Road Service License system that had existed since 1931 was discontinued in favour of allowing operators to start up services as they wished, though the effects often did no favours to the public or the operators in the long run. De-regulation was set for 28th October, but there was no great rush for other services through the Park!

Another local era ceased with the sale that year of the last of the circus equipment from the Smart's Winter

Quarters on a site near the North Street Garage, that land being purchased by HFC Bank to build a new head office. Staff were relocated from Bracknell, with a contract service provided, as we shall duly see, as well as some hires in connection with staff training.

The third era to end was directly a transport one, which started with the death of Alf Moore, founder of *Imperial Bus Service,* the last of the once many Windsor-based independents, based at Clewer Hill and originating in the 1920's. As a result of his passing the sons decided to pack it in at the end of January 1987, so there was more work now available.

Back in May 1986 we have the first mention of a Tachograph, a device fitted to vehicles to monitor the hours of movement and rest etc., all required under the tighter regulations coming into force, and another thing to consider on longer hire jobs where a driver could run out of permitted driving hours. 'EAJ' went to the John Lewis Central Workshop at the Southern Industrial Area in Bracknell for the fitting, whilst that same month 'SNK' was sent to Willowbrook's up in Loughborough for a new windscreen, but the fitter broke the new one, so an overnight stop ensued.

The old Sunninghill Picture House, like many other cinemas, fell on hard times, becoming the Novello Theatre. In this photo 'STL' is getting attention from Brian and Geoff Lovejoy at the bus stop.

There were a number of changes to secondary school arrangements in Windsor during the 1980's, and from September 1986 there was a merger of St. Edward's RC School and the Middle School of Royal Free into the new St. Edward's Royal Free Ecumenical School at Parsonage Lane, just to the west of the town centre, allowing the former sites at Dorset Road and Batchelor's Acre to be re-developed.

However, once the school opened in September it soon became apparent that the old Dorset Road site would need to continue in use temporarily, so the Headmaster Mr. Runyard approached *Imperial* to see if a suitable bus journey could be provided, but in view of the situation with Alf's health, they felt unable to oblige. He then contacted *White Bus,* which already took pupils to all the Windsor schools, and it

was agreed to run on to the new location. However, as the distance was only about a mile, no free transport was granted by the Education Department, so the Head sent a note to parents advising them that the link would start from Tuesday 1st October at a charge of 4 shillings per week, payable to *White Bus.*

The revised Route 02 shows the bus leaving St. Peter's School at Old Windsor, where it had arrived through the Great Park as Route 03 for 8.30, then as Kingsbury Drive 8.31 – Bells of Ouseley 8.34 – The Wheatsheaf 8.36 – Osborne Road 8.41 – Victoria Street 8.43 – Alma Road 8.46 – Spital School 8.49 – Bulkeley Avenue 8.51 – Parsonage Lane 8.55. There was an afternoon link in the same direction, starting from Kingsbury Drive 16.02 via The Bells of Ouseley – The Wheatsheaf – Osborne Road to terminate at the Central Station, after which it did a public service run. In respect of the return run from Parsonage Lane, the bus left at 15.40 and ran as Springfield Road 15.42 – Bulkeley Avenue 15.44 – Spital School 15.46 – Alma Road 15.50 – Victoria Street 15.53 – Kings Road 15.55, to The Wheatsheaf, Old Windsor for 16.00.

The involvement of Local Authorities had also been altered under the Transport Act, whereby they now had responsibilities for assessing public needs and for supporting services more directly. In East Berkshire it also resulted in the tendering of supported services in late 1986. Speaking from personal experiences, the tender process might well produce some savings on budgets, but could sometimes go awry when involving bidders who were yet unproven, some applying for multiple contracts, often without vehicles or even a base locally, an important factor in an expensive area.

So, it can be imagined how Doug Jeatt felt in 1995 when the tender of *Nightingale Coaches* from Maidenhead for the Windsor to Ascot service came out the cheapest and was accepted! It was another aspect of tendering that firms fearful of having no jobs would apply for more than in reality they could handle, and Doug must have been mighty relieved when that operator declined to accept the contract and *White Bus* lived on through yet another crisis!

Staff changes for 1986 saw the addition of Bill Darling and Terry Gibbs, whilst Jim Waterfall had to come off driving duties for health reasons, though would in due course be back but in the office instead.

During 1986/7 the number of schools hiring increased steadily, whilst pre-school groups were sprouting up throughout the catchment area, their more localised jobs such as Beale Park near Pangbourne fitting nicely with other daily commitments. School groups went to the Courage Shire Horse Centre on the A4 west of Maidenhead Thicket, or to Thorpe Park, a new theme park with adventure rides set around an old gravel pit used to construct the M3 motorway near Staines.

Another former gravel working between Winnersh and Hurst was also developed by Wokingham District Council as a watersport centre and nature reserve as Dinton Pastures, and was also set up for school trips.

The first of many outings organised by the Bracknell Jehovah's Witnesses is noted in 1986, some of which took groups to the large regional centre in a former cinema at Dorking. Other diverse groups catered for included the social clubs of Avis Car Hire (Bracknell), DeBeers (the diamond merchants of South Ascot), the Running Horse Golf Society (Bracknell), 3M Social Club, the DEBRA Charity, the Soroptomists (Slough), the King George VI Club (Windsor), John Mowlem Construction (to view its work in London Docklands), the East Berkshire Deanery (Windsor), the civil Service College (Sunningdale) and British Aerospace Social Club (Bracknell).

Datchet Old Folks weekly booking was inherited from *Imperial* in 1987, mostly on a Monday, for many years, with some localised trips and other longer outings to the coast or places of interest. Another enduring seasonal task was to take the Berkshire Young Musicians during July to concerts at various venues throughout the area, all out of school time.

As noted earlier the passing of Alf Moore convinced the family to give up on its circular Local Service for Dedworth and Clewer from Windsor town centre, but in doing so they sold it to *The Bee Line* after the last day on 24th January 1987. That same month *White Bus* had purchased one of the *Imperial* coaches, a YMT-type Bedford (JMJ 633V) which carried a 53-seater Plaxton 'Express' MkIV body, and also featured a wide inward-opening door suited to less able users. It was new in 1980, but would see another 20 years with its new owners, initially as a front-line coach and much travelled, and in later years a daily performer on school contracts, so was a very sound purchase.

Being collected from the Imperial yard at Clewer Hill is Bedford YMT-type JMJ 633V with Doug Jeatt at the wheel, complete with dark curly hair and moustache.

To return to the other activities of 1986/7, covers were provided for *Imperial, Amber & Blue* (Slough) and a couple for *Horseman's* (Reading), as well as providing transport for Griffin Travel of Slough for flights from Heathrow or Gatwick on their holidays. In respect of work for *Winkfield Coaches*, 'EAJ' was loaned to them for 3 days in order that Joe Sutcliffe could take it up to Blackpool, on the trip he regularly organised. On another occasion 'STL' was loaned to *Fernhill,* and Doug was amused to see it on a cover by them of a Charters contract normally run by *Fargo*!

In respect of contracts and route numbers it is worth looking at how these developed over the years. Under the original Charters School needs back in 1958 there had been 4 coaches, seemingly known as 1 to 4, but as time went on they got contract numbers evidently ranging widely. By 1986 those with *White Bus* were 24 and 28, but in 1986 those numbers are prefixed C, presumably for Charters. However, in 1987 they are noted as prefixed M, which highlights the start of the scheme familiar to the author some 15 years later, with a prefix based on the Local Authority in whose area the route originated, there being 6 in Berkshire since the re-organisation of 1974.

So Bracknell Forest BC had B, Royal Borough of Windsor & Maidenhead had M, Newbury BC had N, Reading BC had R, Slough BC had S and Wokingham DC had W, with an S added for Special Needs contracts. When the 6 LA's became Unitary Authorities with the dissolution of Berkshire County Council in 1998, the system continued, though West Berkshire Council replaced the former Newbury BC. At Autumn 1987 the Charters routes were charged as M24 £39.50 per day, M28 at £40, with the late run from Charters at £18.50, whilst the St. Peter's and St. Edward's run was combined at £33.50.

There was also an auction sale at Clewer Hill of buses and equipment on Friday 27th February, and Doug managed to buy a pair of 1974 Bedford YRQ-types with Willowbrook 001-type 45-seater bus bodies (GNH 530N and GNV 983N), both of which had been new to *United Counties* of Northampton. They were in fact very similar to 'STL', and in view of the good relations with the Moore family they were left there, and indeed they were soon joined by 'SNK', which was presumably covering a Charters School contract now also passing to *White Bus,* or to put it strictly in context re-acquired, which had been covered earlier.

Not only did *White Bus* acquire a trio of vehicles from *Imperial,* but also a family member as a driver, a case of buy three, get one free? That was Alan Moore, who was often to be found on the service route, with a ready smile and relaxed manner, who started on 28th April would stay until retirement in May 2000. Also noted occasionally was another of the same family, Roger Moore, on some driving duties during 1987-9.

As soon as it was repainted former Imperial Bedford YRQ-type (GNV 983V) was placed into service, and is seen here passing through Cranbourne Gate onto the public highway towards Ascot. Note how little spare room is available for the bus, so perhaps it is not too surprising that there were some collisions with the gates, at times caused by them moving as the bus approached, and indeed we shall hear of this very bus having such an event. Also note the signs prohibiting the use of the roads by unauthorised motor traffic.

Meanwhile, back in February 1987 Doug Jeatt had attended his first meeting at BCC to discuss a scheme for Concessionary Fares for Senior Citizens, which at that time were determined locally and with varied degrees of subsidy or times of validity. BCC proposed a half-price travel scheme, which was quite typical.

The pair of Bedford saloons from *Imperial* were still over at Clewer Hill when, on the evening of Saturday 2nd May 1987 around 8.30pm vandals set light to the premises. The Fire Brigade managed to pull 'GNV' and 'SNK' out, but 'GNH' could not be saved and was a total loss, without having turned a wheel for its new owner. A photo of the aftermath appears on page 86 in the colour section. Perhaps not wishing to tempt fate with 'GNV', it was quickly repainted white and entered service on 6th May, being lettered 2 days later by Les Clarke.

Amongst the work done involving foreign students was one during July 1987, dubbed 'The Italian Job', so it was a shame that all of the Bedford VAL's that passed through *Winkfield Coaches* had then gone!

Over to the west, *Brimblecombe Bros.* of Wokingham finally gave up even their minibus work in September 1987, another long-established local operator gone.

During the following month came the infamous Great Gale, as Michael Fish will always recall, which brought down many trees affecting services generally, and especially in the Great Park. Drivers used their local knowledge to find ways around blocked roads, and Bill Cathcart and his team were soon out with chainsaws, though many fine ancient trees were lost.

1988 was a relatively uneventful year at *White Bus,* but it is worth taking a quick look at the other operators they regularly worked alongside, such good relations being important for working both ways.

Of the more recent entrants to the world of full-sized PSV's, several had their origins in taxi work, with *Bill Courtney* starting out in 1974 with taxis and then mini buses in Wokingham, and Mark Way on taxi work before starting *Reading & Wokingham Coaches* in 1992. *Hayward's Coaches* over at Caversham was the business started by former *Thames Valley* bus driver Raymond ('Dick') Hayward and his wife Roberta. At Bracknell *Fernhill Coaches* had also been started by another former *TV/Alder Valley* driver Frank Holgate in 1979, joined later by other family members, the name being derived from their Fernhill Close address. *Fargo Travel* was started by Tony Farrugia by 1981, first off at Jameston in Birch Hill, then later at Bay Road, Bullbrook. To the south at Sandhurst *Hodge's Coaches* went back to 1925 and would see Peter Hodge's sons Martin and Paul enter in due course to make the third generation.

Another shot of 'GNV' passing the statue of Queen Victoria which stands by the approach to the castle at Windsor.

Changes in Windsor Town Centre saw the gradual pedestrianisation of Peascod Street, alongside a new shopping development at King Edward Court in line with re-development of the Central Station and Goswell Hill area. The upper part of Peascod Street was at first restricted to limited traffic, but from 9th May 1988 the buses no longer left town that way, so a new stop was placed at its foot in William Street, the *White Bus* service looping around the block to provide a point there for passengers to alight or board.

YRQ-type Bedford (GNV 983V) is seen passing the Duke of Edinburgh pub on the Woodside loop on its way through to Ascot (Heatherwood Hospital).

There were a number of new names in the ranks of drivers for 1988, with Brian Cox from June, Carol Venables and Eric Chambers from July and also from August Ken Acton, though Eric was mainly found on body repairs, just helping out occasionally.

The first of the annual pilgrimages by coach was organised by the Windsor Catholic Association to the Shrine of Our Lady at Walsingham in Norfolk, taking place from Monday 2nd May 1988, when Mick Fazey took 20 passengers on 'JMJ' for 3 days, which included a day out at Sandringham, charged at £550. For 1989 they again booked *White Bus,* thanking 'our own local bus company which looked after us so well', requesting Mick once again, and signing off the letter to Doug with 'love to your Mother'.

Indeed, further regular work was forthcoming with the opening of the large HFC Bank headquarters off of North Street from July 1988. A daily contract brought relocated staff from Bracknell Town Centre and back again, plus on Tuesdays and Fridays there was a run provided into Bracknell and back for shopping etc. A test run was undertaken on 14th July and the service started from Thursday 21st July, with an earlier and later timed run each way daily. Carol Venables with 'EAJ' were the usual allocation to those new duties.

Indeed, several firms in Bracknell soon found the need to take on satellite buildings due to expansion, an example being ICL of Lovelace Road on the Southern Industrial Area, with other offices at Western Road, which *White Bus* often covered transfers for *Fernhill.*

For younger group outings in 1988/9 the Bekonscot Model Village was again popular following a major restoration, along with various local farm centres with lambing or animals to pat. Older school parties went to the new Royalty & Railways Exhibition now in the Central Station, the Dolmech Recorder Factory down at Haslemere in Hampshire, Syon Park (with its free-flying exotic butterflies), Blakes Lock Museum (by the Kennet & Avon Canal in Reading), the Guide Dogs Training Centre at Wokingham and the Hawk Conservatory near Andover as new venues, and Wooden Hill School in Bracknell went off for 4 days in Derbyshire with Mick Fazey aboard 'JMJ'.

Work for St. George's School in Ascot was also most notable from 1989, with a weekly trip to the riding stables at Bearwood near Wokingham, and also the John Nike Ski Centre at Binfield, whilst other trips included Madame Tussaud's and the Planetarium in North London, various theatre trips, and a break to the Isle of Wight via Red Funnel Ferries at Southampton. On one Saturday in February 1989 there were 4 jobs for that school, and by September of that year some 10 jobs a week are shown to a variety of places.

That other private girl's school, Heathfield by the Royal Forester's crossroads continued with a Summer School, with trips booked to various local sporting and entertainment venues, as well as to church.

Work-based social clubs still accounted for regular evening hires, making up for the loss of traditional work for pubs, with Waitrose of Bracknell and Caley's of Windsor, both John Lewis Partnership, had trips to theatres, TV studio shows and to the partner's club at 'Odney' by the River Thames at Cookham.

Other adult groups included the Red Cross Old People's Club, the Bracknell Active Retired, the Blenheim Chapel at Maidenhead, whilst Bob Napper continued to organise trips from the Chalvey area, and Ron Richmond more locally, both usually driving too.

1988/9 saw no fleet changes at all, but in January 1989 a further boost to potential work came when yet another old established coach operator *Carter's of Maidenhead* gave up that side of their business to concentrate on its travel agency instead.

The value of repeat work from satisfied customers can never be underestimated, and Wooden Hill School came back for May 1988 for a 5-day trip to the study centre set up by Berkshire schools at Tirabad in North Wales, with dormitory-type accommodation.

One of the points on the route used at that time to Charters School was under the railway bridge on Dry Arch Road, which gave the shortest access to Charters Road. Here we see Bedford YRQ-type bus (STL 725J) passing under the 8ft 9ins sign, a tight fit through an opening intended for cows when the Staines to Reading line was built. As time went by the types in use precluded the use of the popular short-cut, particularly modern high-floor coaches. Note the blind set to Windsor Great Park despite being on a school contract working.

During 1989 cover work was undertaken also for *Apple Coaches* of Slough, *Courtney Coaches,* and for Thomas International parties arriving at Heathrow and staying at the Burnham Beeches or Royal Berkshire Hotels, including days out locally. An exceptional job in July 1989 saw vehicles required to meet Chelsea Pensioners from the Windsor & Eton Riverside Station for transport to the Royalty & Railways Exhibition at the Central Station, handled in relay.

Examples of costs for coach hire for 1989 were £90 for a 53-seater to Beale Park from Sunningdale, or for Ascot to Weston-super-Mare for £198, whilst the early May day-trip to Spalding Festival of Flowers was £290. On the staff front the additions were Gerry Bourne from April and Ron Coxhead in September.

A certain amount of work for the Licensed Victuallers School at Slough had already come to *White Bus* from *Fernhill Travel* in the form of covers, particularly for day trips, but for the Autumn of 1989 the school was relocated from where it had been established in 1803 to a new 26-acre site at the junction of Fernbank Road and London Road, about a mile west of Ascot. With it came some 700 pupils of both sexes to a mostly fee-paying selective school which included boarders. As part of the relocation those day-pupils attending at Slough were given transport between the two sites until their time at the school was completed, but new entrants could also travel at parent's expense. At first it was *Fernhill Travel* providing that transport, but in due course we shall see how important a move this would become for future *White Bus* developments.

There were further bookings from CAMRA to beer festivals, plus the pre-Christmas outings from local firms, with Technical Indexes and the MET Office from Bracknell choosing the Jack of Newbury pub at Binfield and ES2 the Waterloo Hotel at Crowthorne.

Before continuing with the chronological sequence of events, we must take a look at how the maintenance of the fleet was handled from the late 1970's on. As noted earlier, at first Doug Jeatt did most of the work, as his father had done before him, but of course that meant time when he could not be available to drive or to get on with the myriad other tasks required. So, in the late '70's Maidenhead minicoach operator *Barry Reed* introduced him to Brian Lovejoy, who worked as a self-employed mechanic for 'Bunty' Bennett at his garage at Moneyrow Green.

After that Brian started doing maintenance for *White Bus,* with buses usually taken over to him and returned later in the day, the distance being only a few miles. After Bennett sold up, Brian continued to work in his own right, first at Silverdale Farm in the Drift Road before joining a haulage firm at Prior's Garage in Priors Way, Maidenhead, though freelance still, again with vehicles taken over. That site was re-developed around 1982, about the same time that the workshops at North Street Garage were extended, so he came over to the site, sometimes doing other customer's vehicles there, but much handier for Doug.

1990-2 The Re-unification

The year 1990 will be recalled in history as when the divided parts of Germany were once again re-united after the Berlin Wall came down in that city. Nearer to home, local history was made when Dick Mauler sold *Winkfield Coaches* to Doug Jeatt later in that year, rendering all operations from the North Street site under the one ownership after 35 years.

The year started with a bang, with 'GNV' hit by a car at Crimp Hill on 11th January, its driver being lucky to survive, as the bus front axle needed to be removed.

The damaged 'GNV' is seen at North Street being repaired by Eric Chambers, who handled most of that type of work over many years, and also occasional driving, and it returned to service on 6th February.

A bit more rail-replacement followed on 25th January 1990, albeit impromptu, when trees blocked the line between Ascot and Aldershot, *White Bus* stepping in for 5 hours, with some spare capacity that day when St. George's School cancelled some planned trips due to weather conditions.

On the daily contracts from March the HFC Bank runs still left Bracknell Town Centre at 7.15 and 8.15, but now featured return trips at 16.10, 17.10 and 18.10 to accommodate flexible working. Incidentally, it was the vehicle doing school swimming that usually also covered such workings.

Mention has been made of the Concessionary Fares Scheme for Over-60's, which was also applicable for Registered Disabled Persons, which Bracknell Forest BC and the Royal Borough of Windsor & Maidenhead introduced from April 1990. As a result there was a 50% increase in ridership in the first week, a pleasing boost to the bus service. However, it must be said that such schemes were locally determined, so operators might have to consider different conditions emanating from the Local Authorities within their area.

Evidently *Sky Valley Cars* had been pleased with the service from *White Bus*, as it subbed out several jobs to the firm in 1990, one with corporate clients collected from London City Airport to stay at the Kensington Palace Hotel, with a day visit to Radlett, followed by a night out at a London theatre after that. Another London-based job saw an airport transfer and 4 days of hops between London hospitals. The third example saw a party of business guests from Japan taken to various sites, with Heathrow flights, several of these tasks reflecting inward investment of the day.

During 1990 the amount of work for the Licensed Victualler's School grew steadily, some just local to sporting fixtures and others further afield. There were also jobs for Eton College, subbed from *Windsorian,* and Lilly Research at Erlswood Manor, Windlesham gave their annual seaside outing to the Company.

Even more pre-school groups had been formed, so the early months saw pantomimes attended locally, whilst the lambing season was very popular, as was Beale Park. Older children visited the Cabinet War Rooms in Whitehall, the Imperial War Museum, Kennington, Mapledurham House on the River Thames, Marwell Zoo and the Cotswold Wildlife Park up at Burford in Oxfordshire. In addition to the annual pilgrimage to Walsingham, there was also a day-hire requiring a 53-seater plus a 23-seater from *Fernhill Travel.*

The Summer school holiday saw continued bookings for transporting foreign students, with a large party staying at Royal Holloway College at Englefield Green, as well Heathfield School requiring a vehicle on a daily basis during its Summer School again.

This aerial photo from the White Bus Archive gives an opportunity to study the development of the site at North Street. The original shed towards the front has been improved and re-roofed, whilst behind that is the garage extension, with offices to its right. The brown-painted Mercedes scout bus is in front of the office.

105

Another view of Bedford YRQ-type (GNV 983N) shows it by the pink-painted gatehouses at the end of the long drive leading to Royal Lodge. The bodywork by Willowbrook was quite functional but well built.

Other adult groups hiring in 1990 included the busy Waitrose Social Club, with events as diverse as Level 42 playing at Hammersmith and Blind Date live at TV studios, whilst CAMRA went to Midhurst for the beer and both Express Dairies and a party arranged by the Belvedere Arms at Blacknest went to see the London Lights before Christmas, each finishing off with the seemingly obligatory fish'n'chip supper.

It is also notable that by August 1990 *White Bus* was covering the Rathdown Industries contract for Dick Mauler very frequently, as well as a shopping run into Windsor which had taken locals from spots not on the public bus routes for many years a couple of times a week for a few hours. At other times *White Bus* was covering their obligations by vehicle but using driver Les Spong who worked for *Winkfield Coaches*.

All of this was of course leading up to the retirement of Dick Mauler, then in his 75th year, in December 1990. Understandably, Doug Jeatt was offered the chance to buy him out and, as has already been noted, there must have been many who didn't even know the concerns were operating separately. The only outward sign of the acquisition was that two of the coaches were duly repainted white and re-lettered, whereas the Maulers had mostly operated their coaches in the livery of former owners. Details of *Winkfield Coaches* activities and fleet will be found on pages 151-155.

In respect of the vehicles transferred from *Winkfield Coaches* to *White Bus,* all 3 were Bedfords, comprising a 1975 YRT-type with 53-seater Plaxton 'Panorama Elite' MkIII body (GHD 668N), a 1975 YRQ-type with similar body seating 45 (KDT 281P) and a 1977 YMT-type with Duple 'Dominant' 53-seat body (RLW 778R).

The vehicles were transferred during December 1990, whilst contracts and licensing matters were concluded in March 1991. The liveries of the trio of coaches can be seen in the photos on page 114, whilst that carried by 'GHD' gave rise to the nickname as the 'Liquorice Allsort' after a brand of sweets of the time. In fact that vehicle saw only a little use before being laid up in the garage, after which it was used for parts until finally going for scrap in 2003.

Bedford YRQ-type (KDT 281P) is seen after re-paint into the white livery and outside the Hernes Oak pub in North Street. It proved a useful addition and was used for a further 13 years, and it survives in preservation with Elderson Coaches of Bromsgrove.

'RLW' was also repainted and saw a further decade in service, whilst drivers Joe Sutcliffe and Les Spong also transferred to the new owner, both as full-timers, though Dick Mauler now gave up PSV driving.

All of these Y-series Bedfords had either the 466cu. ins. or 500cu. ins. engine, and after the re-unification the 3 mentioned above worked alongside the sextet of Bedfords which were the 1971 YRQ (STL 725J), the 1974 YRQ-types (SNK 255N and GNV 983N), the 1979 YLQ (HRO 958V), 1979 YMT-type (EAJ 327V) and the 1980 YMT (JMJ 633V).

Weekend rail replacement work was a useful addition, with vehicles free from weekday commitments such as school contracts. Here we see Bedford YMT-type (RLW 778R) in the white livery and outside Langley Station on that task. At such times there were often part-time drivers available including those who were enthusiasts wishing to keep their PSV Licenses going in order to drive preserved vehicles at other times, but it involved waiting around.

Also taken over from *Winkfield Coaches* were the daily commitments including school and work-related contracts, as well as a busy private hire diary. Apart from the Rathdown run there was another Charters run now added, which had its origins in the opening of the school back in 1958, though obviously the routes had been varied over the years to accommodate changes in intake and content of year-groups.

Joe Sutcliffe had for many years organised an outing to Blackpool for several days to see the illuminations, which continued, whilst the coming of Les Spong into the fold saw him as the usual driver for the Bingo run from the Great Park to Windsor's Regal Cinema, as he lived at Mezel Hill, taking the vehicle home with him.

Work for HFC Bank had also increased, as apart from the contract runs there was an increased instance of extra transfers for staff attending training courses from the Hilton Hotel at Bracknell or the stations at Reading and Slough from March 1991 onwards.

Having more vehicles available also led to increased covers for others, a notable one being the Cox Green School contract for *MD Coaches*, which led to some private hires from that school and, ultimately, to the firm becoming the contractor for the daily route.

Amongst the local hires, Mrs. Probert of Cheapside made regular bookings, whilst the tireless Maureen Fitzgerald continued to organise for the York Club. A large party from Windsor Safari Park took a pair of 53-seaters to Chessington World of Adventure for a visit, whilst the Met Office Theatre Group went up to London venues monthly, and Slough Naval Club went down to Portsmouth, Chatham and London for various commemorative events during 1991. Foreign visitors continued to be carried, with a Czech party picked up from RAF Manston, 35 Portugese flew in to Heathrow for transfer to the Runnymede Hotel, whilst the Japanese girls coming in for the Heathfield Summer School needed a luggage van as well!

On the St. George's work, a weekly run for rowing on the Thames at Staines started with the Spring Term 1991, and in general the jobs for that school built up. There remained regular transfers for Scouts and Guides to their camps, either as local as Stonor, north of Henley, or to Hamble and the New Forest, whilst the Windsor Sea Scouts headed inland to Sandhurst and the Royal Military Academy, though it does have a large lake! A somewhat larger event saw 359 Cubs and Beavers and their helpers taken from the Windsor and Ascot area to Poole for a gathering on Brownsea Island, birthplace of Baden-Powell's scouting movement on Sunday 9th June 1991 involving 'EAJ, JMJ, RLW and SNK', plus a 53-seater from *Fernhill* and a pair from *Tent-Tours*.

The frequent pick-ups from St. George's School at Ascot were managed from a hard-standing area built below the school site, where 'EAJ' awaits in the sun for girls to come down from the extensive site above.

The Rathdown Industries contract had been running for many years, bringing workers in from outlying areas to the light engineering works at London Road in Ascot, but the firm had decided to take up an offer to relocate to the expanding Wiltshire town of Swindon. Some of the staff would also be transferred, with re-housing also included in a New Town in all but the name, so several hires took place to take staff to view the new premises and area, but the daily job ceased on Friday 27th September 1991.

Bedford YRQ-type (KDT 281P) had the MkIII version of the 'Panorama Elite' body and continued with the relatively low seating capacity of 45. It is seen in the depths of the Great Park near Norfolk Farm on a Route 13 school journey from Windsor to Ascot.

On the service route from Monday 28th October 1991 the running order in the Sunninghill and Sunningdale area was amended, with journeys reaching Cheapside (either through the Park or via the main road), then as Sunningdale (Broomhall Lane) – Sunningdale (The Rise) – Sunningdale (School) – Sunninghill (School) – South Ascot – Ascot (Station) – Ascot (Horse & Groom) – Heatherwood Hospital. Exact stops varied as before, with certain journeys direct from Cannon Corner to Ascot High Street. In the new travel guide by RBWM the route is shown as 'A', though known at *White Bus* from at least 1986 as '01', both of course highlighting it as the first.

At that point *The Beeline* also had services 187/197 from Bracknell, which covered the section between Heatherwood and Broomhall Lane, as that stop made a convenient terminus off the busy A30 but close to Sunningdale Station, with certain journeys to Charters School at appropriate times. However, those services had been reduced and had led to RBWM seeking the changes in the *White Bus* route. As a result of the changes buses no longer called at Blacknest Gate, but it was stated on the timetable that passengers desiring that point could contact the office to arrange a deviation.

As often happens in the life of such operators, when one door closes another opens. Now with a coach spare from its Rathdown run, *White Bus* answered the call from Dick Hayward to cover his afternoon run from St. Crispins School in Wokingham taking pupils to California Crossroads. The route was W39 of Wokingham DC's school contracts and, after a week or so covering the afternoon run for *Hayward's* it was decided to offer *White Bus* the operation from Monday 25th November 1991, plus the opportunity to tender when reviewed next. As such a service is still run by the Company, it is clear that being in the right place at the right time affected such fortunes.

'RLW' passes the ornate Victorian Wokingham Town Hall on the afternoon run of W39 from St. Crispin's School. For many years a bell was rung from the tower of the Town Hall each evening to guide any travellers lost in Windsor Forest, the whole area once being within its boundaries.

No fleet changes took place in 1991, so the 9-strong contingent of Bedfords was kept busy. There were even some evening 'drives' arranged, along with tours of the Great Park or nearby Chilterns for the Autumn colours, one of which saw the Royal Household out in the Marlow area. School visits now included the Didcot Railway Centre, home of the Great Western Society keeping steam alive, whilst in December 1991 local churches Windsor URC and St. Francis RC put aside any detail differences and hired a coach down to Portsmouth Cathedral. The Crown Estates staff were taken to their carol service at the Royal Chapel shortly before Christmas, whilst corporate festivities took the fleet to Chico's on the A30 at Virginia Water for Novel, to the Jade Fountain for ES2, and the larger Met Office was split between The Cricketers east of Warfield and the Admiral Cunningham in Priestwood, Bracknell.

The various gate-houses in the Great Park vary in their style and size, some having been enlarged over the years to their present situation. However, that at Prince Consort Gate was a quite grand structure from the outset when built in 1862. Here we see Bedford YRQ-type (SNK 255N) on its way to Windsor after passing through the gate. This vehicle had a body suited equally to use on the bus service and for private hires. The windscreen offered good visibility but was prone to damage – Doug still has a spare one!

One driver is noted from November 1991, when Ken Cooper joined, actually being brother-in-law to Alan Moore, as they had married twins. He was also the District Scout Commissioner, and another to gain his PSV License through the firm! Doug kept in touch with many former drivers to use when required.

For some years Doug Jeatt was involved with the committee for the 1st Cranbourne 11th Windsor Scouts, his sons of course being members, so it is perhaps unsurprising that the subject of a scout bus should be addressed. Dave Empson and George Lynham were the Scout Leaders in the early 1990's, and both were PSV trained at *White Bus* in connection with those duties, after which both also undertook work for them. Dave started to appear on the books from April 1992, though he also had his own courier's business under the names Dawe Transport, whereas George worked for Barclays Bank, and later found a ready place at Winkfield when that ceased.

Details of the initial scout bus are incomplete, but it was a brown-painted Mercedes of medium size and nicknamed 'Scooby Do'. It was maintained by *White Bus,* but kept at St. Peter's Hall in Hatchet Lane in Cranbourne, the place where the Scouts met, and one night it vanished, believed taken for scrap by Travellers, and was never found again! Next to come as its replacement was a former British Airways crew-bus, with a Dennis G10 chassis and Wadham-Stringer body (C911 VLB), new in 1986 and acquired in 1996, a photo of which appears on page 94. It proved a bit troublesome, boiling the radiator on long runs, after which the Leaders instead hired a vehicle as self-drive from the firm.

To return to developments with the LVS at Ascot, *White Bus* started to cover the morning run from Slough to the school from Wednesday 30th January 1991, being placed somewhat nearer that *Fernhill.* The move certainly increased the hires from there and, like most foot-in-the-door situations would result in even more daily involvement with that establishment. There were, however, no changes to the fleet in 1992.

Vehicles were hired for taking extra Police Officers to Ascot for Royal Ascot Race Week in June for traffic control, whilst in those days it should be recalled that the old course also led to a road closure at times. As the New Mile course ran across the Winkfield Road just near its junction with London Road, it was necessary to fill the gap when racing was in progress. Each year the trucks of Harry A. Coff brought sand etc. for the purpose, before turf was laid on top, the road later being cleared, but these days the new underpass has removed such distruption!

Also in 1992 the York Club celebrated 50 years since its opening, with guests brought in from every corner of the Great Park by the firm's vehicles. The Royal Household Social Club went to the Royal Tournament at Earls Court, whilst Waitrose SC saw Eric Clapton and Diana Ross at Hammersmith, and other parties booked for Torville & Dean on Ice at Wembley. Two coaches were needed in May 1992 for the Spalding Flower Festival, whilst the Woodley Labour Party organised a trip to see the recreation of the GWR in the Vale of the White Horse in miniature at the Pendon Museum, Long Wittenham near Didcot.

South of Wokingham and near the Crooked Billet at Gardeners Green is this ford through a tributary of the Emm Brook, with 'JMJ' taking a short cut back from a rally at the TRRL site at nearby Crowthorne.

The pre-school groups continued to grow, and on one day there were 3 *White Bus* vehicles at Finkley Down Farm, just north of Andover for the lambing season. School work had continued to grow, with 34 different schools noted during 1992, ranging from as far apart as Egham to the east, Whiteknights in the west and Wexham to the north. New venues visited included Broadmoor Hospital, RAF Museum Hendon, Loseley Park near Shalford and the Mary Rose in Portsmouth.

During the school breaks, the fleet was again busy with foreign students, one party collected on behalf of *Tent-Tours* at Easter from the Harwich ferry terminal in Essex, whilst *Impact Coaches* asked for 3x53-seaters from Winchester to Royal Holloway College in the Summer and day excursions to Oxford, Brecon, Clacton and Stratford-upon-Avon. More locally the girls of St. George's, Ascot added a weekly trip to Heathlands Riding Centre, very close to the ford seen on the previous page with 'JMJ'.

Groups for older people also expanded in number, with the Forget-Me-Not Clubs of both North and South Ascot enjoying regular outings, along with Old Windsor and Sunninghill Darby & Joan Clubs and the Chavey Down Pensioners, all extending the age range catered for by the Company. Local churches also were well represented, including the newer arrivals such as the Kings Church in Windsor and the Kerith Centre at Bracknell, whilst Jehovah's Witnesses still featured.

The Crispin pub runs through White Bus history like the name in a stick of rock, but on this occasion we see Plaxton 'Supreme IV'-bodied Bedford YMT-type (EAJ 327V) on a private hire involving that hostelry.

Shortly after the change of terminal point Tony went up onto Windsor Castle to catch 'SNK' parked at the stop opposite the historic Theatre Royal.

Incoming buses ran into town from Kings Road, then Sheet Street – High Street – Thames Street, and around the triangle formed by River Street – Thames Avenue to reach the terminus. On leaving the bus then went down River Street again before continuing as Goswells Road – (lower) Peascod Street – Victoria Street – Sheet Street – Kings Road.

Involvement with the LVS increased during 1992 in respect of daily runs, and it is worth noting that one of the staff not to relocate to the new site was none other than Jeanette Wright, wife of Tony, who had been PA to the Head – small world, as they say.

Although the other bus services operated by *Beeline* had moved from the Central Station at Windsor, the *White Bus* had stayed on, with Madame Tussaud's constructing a new turning circle in the old car park. However, as the King Edward Court development in the area between Peascod Street and Goswell Hill took shape, the use of the station for retail also became more intensive, finally ousted the bus service from Monday 13th July 1992, to a new terminus opposite the Theatre Royal down on Thames Street.

The timetable changes of 13th July also saw the firm obliging the school with an afternoon link across to Ascot – South Ascot – Sunningdale as counterpart to the 07 morning run from those points. However, for some reason it was shown on the Route 01 table as School Days Only and starting from Bracknell Bus Station, though the vehicle did actually carry a board bearing '07 Sunningdale'. Departing Bracknell at 16.00, it was regularly used by 3 ladies for the journey to Sunninghill, and although the link was kept LVS discontinued it from early 2015.

The afternoon Bracknell –Ascot – Sunningdale link by Route 07 was worked off the back of the St. Crispin's School California Crossroads run, and here we see 'RLW' pulling into the Bus Station by the Market Inn.

Another arrangement had originally been discussed by LVS with *White Bus* as a run starting at Egham and proceeding as Staines – Thorpe Green – Virginia Water – Sunningdale – South Ascot – Ascot Station, but did not materialise as such because pupils coming from most of the easterly points could instead use the train to Sunningdale and then the bus inwards which started on 3rd September 1992. It was duly cut back to operate between Ascot Station and the school from September 2012, with casual fares of £1.50 per single trip of just 10 minutes, still known as Route 07 and fitting in nicely with Charters School commitments.

The original Slough to Ascot operation had gone to *Fernhill,* but from 3rd September 1992 a second homeward journey at 17.10 was added to the existing 16.10 run, becoming registered by *White Bus* at that point, with those pupils not part of the relocation paid for by their parents direct to the Company. From that same date it was known as Route 08, though confusingly it would be the first of three separate ones to bear that designation, whilst transport from Slough and Windsor would be incorporated into the service starting in The Chalfonts as Route 06 in due course.

It should also be noted that LVS also operated some routes with its own minibuses, whilst although *White Bus* usually started runs for that school on a contracted basis, it duly registered the operations as that meant that the fuel-duty relief could be claimed.

Transport for company Christmas dinners certainly declined from now on, like the pub darts had done so before, but it is worth noting that whereas in December 1991 *White Bus* had covered the event for the publisher McGraw-Hill, based near Maidenhead on behalf of *Tent-Tours,* for 1992 it approached the Winkfield-based operator direct. Such opportunities were what built the private hire up over the years.

1993-5 An Enthusiast's Group

In general 1993 was a relatively quiet year for changes, both on services and the fleet. The only event on the latter was finally letting 'GHD' go during April when, after being stripped of useful parts, it was dragged out from the shed and towed away by Warfield Garage to Blackbushe for scrap (see p.93).

Contracts to Charters School were increased that year due to further intake from Bracknell, including a run starting from Martin's Heron, and though none came direct to *White Bus* covers are noted for *Courtney's, Fargo* and *Fernhill* at times. There were a number of reasons for operators seeking assistance, a shortage of drivers or vehicles or a late-running excursion.

Here we see 'HRO' turning into River Street on its way to the terminus opposite the Theatre Royal.

As for school outings in 1993, the girls of St. Georges School in Ascot flew off to Lyon in February, whilst the party from Pinkneys Green School went to the Isle of Wight for 5 days in May on a field course. Other venues popular then were Butser Iron Age Village, the re-opened Warwick Castle (now Madame Tussaud's), the Thames Barrier, whilst one party went to Bristol Docks to view the replica of the 'Santa Maria' which had been constructed to commemorate the voyage of Christopher Columbus 500 years earlier in 1492.

The charity DEBRA regularly hired to take children with life-reducing skin conditions on outings, whilst their parents attended a conference, and for that task 'JMJ' was chosen for its wide entrance, always taken by Ron Richmond. For younger groups Trilakes near Yateley was a good choice, with its tame animals and picnic tables by the flooded former gravel workings.

Amongst the adult outings were those connected with rugby, one for Wndsor RC to a match down in Dorset, which probably led to the firm being favoured to take Rosyth RC from Kings Cross Station and onto matches for 3 days in the local area. The British Airways Band used *White Bus* for various events, and Newbold College (Latter Day Saints) at Binfield chose them for a tour of the Great Park, the year ending with the Coldstream Guards going to London to see the Christmas Lights with a fish'n'chip supper.

Also of particular local significance was that part of the Great Park hosted a large encampment under the title WINGS (Windsor International Guide & Scout) Festival, which returns every 5 years or so, with the firm inevitably taking various more local groups to the site in the north of the Park.

Another significant 'arrival' in April 1993 was not entirely new to the North Street site, being Geoff Lovejoy, who as son of Brian had been coming to the yard for a number of years. That had decided him to work for *White Bus,* and in the meantime he gained his Driver's License for PSV's with the *Beeline,* finally getting the job he always wanted at Winkfield.

There were no changes to the fleet during 1994, but on the staff side Chris Empson, Graham Messenger and Chris Barber were added to the drivers, whilst it is noticeable that Doug Jeatt now rarely featured in the list of scheduled duties, taking more time to manage.

Once Geoff got settled in, he approached Doug about setting up a White Bus Services Enthusiast's Group, which was supported by the loan of a vehicle for outings to rallies. The initial event was in April 1994 to the 'Cobham Rally', which that year was held at Apps Court Farm, Walton-on-Thames (see page 94). After that a *White Bus* featured at many events in the South, along with some further afield such as Duxford and Warminster, whilst the Group produced fleet lists and potted histories and handed out timetables.

By the Summer of 1994 the school operations in the RBWM area were shown in the Area Travel Guide, and *White Bus Services* were then operating as below-

Route 02 – St. Peter's School, Old Windsor – The Wheatsheaf – Osborne Road, Windsor – Spital School – Bulkeley Avenue – St. Edward's School/St. Francis RC School, Parsonage Lane, the return journey of which saw the bus run back to Windsor in service by the direct route of Osborne Road to Theatre Royal in order to position for a public journey.

Route 03 – Winkfield (Hernes Oak) – The Village – Royal Lodge – St. Peter's School, Old Windsor, with a return journey tabled on Route 01 as 15.30 starting from Theatre Royal.

Route 13 – Winkfield (North Street) – The Avenue – Shepherd White's Corner – Cheapside – Norfolk Farm (Great Park) – Cumberland Lodge – Chaplain's Lodge – The Village – Rangers Gate – Windsor (Bolton Road) – Victoria Street – Parsonage Lane – Hatch Lane (for Clewer Green School). The home run called at Norfolk Farm, Ascot Gate, Blacknest Gate and Cannon Crossroads by request, before continuing onto Sunningdale (Charters School) to form the 16.30 'late coach', which followed a similar route to the 24 other than omitting Heatherwood Hospital, but could also be requested to run on to Crouch Lane, Winkfield Lane and the Drift Road. Also note that Route 03 and 13 could exchange passengers at The Village to reach their appropriate destinations, as seen on page 133.

Route 24 – Winkfield (Hernes Oak) – North Ascot (New Road) - Fernbank Road (Royal Hunt) – London Road and Fernbank Road junction – Heatherwood Hospital – Ascot – South Ascot – Charters School.

Route 28 – Winkfield (The Squirrel) – Crouch Lane – Cranbourne (Lovel Road) – North Ascot (The Avenue) – Woodside – Cheapside – Cannon Crossroads – Charters School, and by request would call at Blacknest Gate.

Mention has been made of loans between local operators, and here we see 1978 Bedford YMT with Plaxton 'Supreme' MkII body (YPB 836T) in the North Street shed during May 1994, with a board for Fernhill's Route 3 to Charters School.

   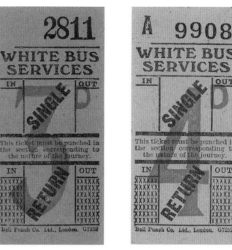

  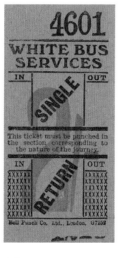  

White Bus used Bell Punch pre-printed tickets for many years, with some additional return values from another un-credited source, and shown here are a selection from those still held in the White Bus archives and therefore in un-clipped condition.

113

Top – We start this second colour gallery with a page dedicated to the Winkfield Coaches fleet, which operated quite separately from 1955 to 1990, albeit from the same North Street premises. 23 coaches went through that fleet and most ran in the livery of their former owners. GHD 668N was a Bedford YRT with 53-seater 'Panorama Elite' MkIII body by Plaxton and new in 1975. It came to the fleet in 1985, in fact the last coach to be acquired, and is seen at the North Street garage.

Middle – This January 1987 view shows the other two coaches which would survive with 'GHD' to be taken over by White Bus in 1990. KDT 281P was a Bedford YRQ-type with Plaxton 45-seat body of 1975, whilst RLW 778R was a 1977 YMT-type with Duple 'Dominant' 53-seater body which arrived in 1981. Some vehicles carried Winkfield Coaches, whilst others had VM&E (or V&E) Mauler for the man-and-wife partners Viv and 'Dick', covering coach work such as private hire and school or works runs.

Bottom - Here we see 'RLW' again in September 1990, shortly before Winkfield Coaches was re-united with White Bus, and in this view it is actually on hire to them! It would duly receive full WBS livery and see use for a further decade. The fuel pumps can clearly be seen to the right, whilst the old brick-and-timber garage built many years before is showing signs of disrepair. The yard area was made up now, and the White Bus vehicles parked in the open to the right as seen above.

Top – One of the biggest boost to White Bus school operations came with the relocation of the Licensed Victuallers School from Slough in 1989, which moved to a new site on the corner of Fernbank Road and London Road in North Ascot, and just some 3.7 miles from North Street. A number of routes have been developed to bring day students in from as far away as The Chalfonts and Camberley. The school also generates private hire work, and here we see 'RLW' and 'WRT' leaving the school.

Middle – Many locals have used the bus service for many years, by traveling to school, then to work or as part of a family. The Jeatt family has put in its three generations of loyalty to its passengers, through often very lean times, which in turn has been repaid by both staff and locals. Here we see long-serving driver Alan Moore with two of 'the regulars'. It is difficult to see how many of those along the route would have fared had it not continued.

Bottom – For an operator with weekday commitments weekend rail replacement work is a useful income and can be covered with part-time drivers. The White Bus vehicles could be seen on such work along the Slough and Windsor lines, from Maidenhead to Bourne End or Marlow, as well as those out from London Waterloo towards Guildford, Reading and Weybridge. Here in 1998 we see Doug Jeatt at the wheel of hired Plaxton-bodied Leyland Tiger HIL 4017 of Edward Thomas & Son of West Ewell at Virginia Water Station.

Top – We have seen how seating capacities had risen over the years from the original 14-seaters, with 20, 26, 29, 41, 45, 53 and even a 63-seater by 1995. Much of that was aimed at economies of scale on contracts, but that left smaller private hires as unviable. So in July 1997 this 1986 Iveco 49.10 with Robin Hood 'City'-type 21-seater body was obtained. D473 WPM is seen outside the Hernes Oak pub and opposite the North Street garage after full repaint.

Middle – As well as small party hire, 'WPM' was also used on some lighter journeys, and on this occasion is seen at Bishops Gate heading for Old Windsor. Just outside the Park and to the left is the Fox & Hounds, a popular place for lunch or a stop off on a walk to nearby Cow Pond or even through to Windsor via the Copper Horse and Long Walk. Note the addition to the traditional lamps of high-mounted security cameras, illustrating the heightened security around the Royal Family due to events at the time.

Bottom – In this final view of little 'WPM' we see it leaving the Park through Cranbourne Gate, with its pink-painted lodge. It is on Route 01, but was only kept for 3 years, departing the fleet in April 2000, without another small-size vehicle coming as a replacement. It did stay fairly local as it passed to Mark Way of Reading & Wokingham Coaches, the two concerns often working together on larger hire jobs. Note the large oak by the junction of the entrance and the public highway.

Top – *Hires involving the girls of the Ascot-based St. George's School feature throughout this page, a task with which White Bus has been favoured for many years, both locally and further afield. Bedford YNT-type JNM 747Y was new in 1983 and carried a Plaxton 'Paramount' MkI-style 53-seater body. It was acquired from Grey's of Ely in 1997 and is seen here still in the full livery of its former owner when pressed into use in December of that year, and it is passing down the lane beside the school building.*

Middle – *Just before the end of the Winter term each year the girls went to Windsor Parish Church for a carol service prior to breaking for the holiday, a job involving some half-a-dozen vehicles. On this occasion in 1998 we see coach 5 as B542 OJF, a 1985 Bedford YNT-type with Duple 'Laser' 53-seat body, along with coach 3 as EAJ 327V, a YMT-type new in 1979 and with a Paxton 'Supreme' MkIV body with 53 seats. In this case it was also necessary to hire in two coaches from Fernhill Travel, Bracknell.*

Bottom – *Further afield we see Bedford YNT-type 53-seater B30 MSF carrying a Duple 'Laser' MkII body, a fine purchase in 1998 when 13 years old. It is on the extended trip to Slapton Sands in Devon, where the unique natural history can be studied. Taken in June 1999 with Mick Fazey at the wheel, the coach is on the narrow coastal road between the sea and the water of Slapton Ley, whilst just out of shot is a memorial to the US troops trained there for D-Day.*

Top – *The theme for this page is inherited colours, and here we see JNM 747Y after it received White Bus Services fleetnames, but was still in the green and cream scheme as acquired. It is seen at Langley Leisure Centre in Parlaunt Road in October 2005 on a regular transfer of students using the facilities there. It was sold 13 months later and therefore never had a full repaint. Operators who already served schools have a natural advantage in quoting for such in-between jobs.*

Middle – *Bedford YNT-type B542 OJF was featured on the previous page, and here we see the interior view looking to the rear. The seats of this 1985 Duple 'Laser' body are both comfortable and lack the dust-traps of earlier designs – compare this with the interior of 1980 Paxton-bodied JMJ 633V on page 85, showing how coach bodies continued to be refined. This coach put in 9 years with White Bus, and this photo was taken at its withdrawal in May 2006, a credit to how well it was cared for.*

Bottom – *Another coach running in acquired livery was 1987 D259 FRW, a Bedford YNV-type, which had a Duple 320-style body and was acquired in 1999. It had quite a few previous owners, and is seen arriving at the Route 01 terminus at Heatherwood Hospital, before circling the roundabout to pick up at the shelter seen to the left. The return of White Bus to that spot echoes the earlier service of Ackroyd, when the buses were kept at the Royal Ascot Hotel, which stood close by.*

Top – The next coach to be acquired was another YNV Bedford chassis, but with a 55-seat 'Paramount' MkIII body by Plaxton. It was new in 1987 as E849 AAO and came 12 years later to join a fleet which, apart from the 21-seater Iveco, was entirely made up of Bedford models. Indeed, the first example of that make had entered the White Bus fleet in April 1948, whilst the final one of that make was disposed of in June 2008, giving a continuous presence of the manufacturer for over 60 years! 32 examples passed through WBS, whilst a further 15 featured within Winkfield Coaches.

Middle – Former Fernhill Travel coach TIW 2795 was the final Bedford to be acquired, and was a YNV-type with Belgian-built Van Hool 'Alizee' 53-seater body. New in 1986 it had run under another three registrations before being given this Irish one, a once common way of disguising the age of coaches. It was duly repainted as seen here opposite the Crown-built Queen Anne's Cottages on Kings Road, allocated to former estate workers. The coach awaits a private hire from the residents.

Bottom – Another view of 'TIW' shows the offside of the Van Hool body, which was the first non-British example for White Bus, the European builders coming into the UK market during the 1980's. This shot was taken in August 2002 in the Great Park, whilst the coach stayed until April 2005. Note the mown field behind, whilst the gnarled old oak tree had either been hit by lightning or damaged in a gale.

119

Top – Production of bus chassis by Bedford ceased in 1986, after which White Bus ran its examples as long as was practical. For a new generation of buses two Dutch-built DAF chassis with Optate 'Delta' 49-seater bodies were bought secondhand in April 2003. R89 GNW is seen at roughly the same spot as one of the Dennis Aces of the 1930's outside The Crispin, with Lovel Lane off to the right, which was used on certain journeys via Fernhill and Hatchet Lane. These were the first low-floor types with spaces for buggies or wheelchair users.

Middle – 'GNW' is again seen in Windsor in 2005, by which time the Borough Bus lettering had been added in royal purple. The bus is heading for Ascot. Even Windsor, with its well known historic centre, was not immune from examples of unattractive office blocks, which usually replaced former residential properties. The DAF buses proved to be very reliable, and increased seating on the service route.

Bottom – A shot that was almost bound to occur, has George III riding on the roof of the other DAF/Optare bus S158 JUA as it passes the Long Walk with the Copper Horse in the background. This bus had been new in September 1998 and was on Heathrow car park shuttles. It carries the Borough Bus lettering, along with places served along the roof, reflecting the increased role of Local Authorities in reviewing the provision of essential bus services in response to legislation in 1970's-80's.

Top – *The Optare-bodied DAF DE02's were the usual performers on Route 01, and here we see 'JUA' in the first of a trio of daily locations. It is seen at The Village post office stores as it heads for Windsor. They had nice wide entrances and good visibility for the driver and the passengers alike. Note the use of yellow lettering for the destination blinds at that time. Their low-floor was called for by the Royal Borough of Windsor & Maidenhead in the process of grant aid.*

Middle – *The same bus is seen at the Windsor stand opposite the Parish Church in July 2003, before the addition of Borough Bus branding, and is preparing for the return journey to Ascot via the Royal Lodge. Note the flags for the Windsor Festival, and the roadway was part of the route used daily for the popular changing-of-the-guard parade between the Castle and the Victoria Barracks. The original White Bus of Ackroyd started from close by the Parish Church, just east of the Guildhall to the left.*

Bottom – *In order to run to Ascot from the above stop, buses turned in front of the Castle, then right again into St. Albans Street, and past the Royal Mews to re-emerge at the junction of the High Street and Park Street, a feature still used if the normal route is closed for road works or when buses do not proceed further east on local routes. A general reduction in car access has improved congestion in that area, though buses can still use the road.*

Top – The initial DAF's were joined by a further used example in January 2004, which had worked at Luton Airport since new in 1997. It was registered P131 RWR in Yorkshire like the other examples, the Optare works being in Leeds. It is seen in April 2004 passing the playing fields of the York Club, opened in 1951 and seen in the background. That was the focus for the many social and sporting groups set up for the community of the Great Park, and the source of many private hire jobs for White Bus.

Middle – There were a number of phases in the pedestrianisation of the Peascod Street area. 'RWR' is seen after the stretch from the Castle to the south was closed off to all traffic, and it is shown coming up from Victoria Road, past the new bus shelter, to turn right into William Street in 2006. The Windsor Relief Road built in the 1960's and the closure of Eton Bridge took away most of the traffic that had once passed through the town.

Bottom – The next coach to be purchased was an Iveco 'Euro Rider' 35-type, which carried a Spanish-built Beulas 'Stergo E' 49-seater body, complete with toilet compartment. New in 2000 it came from Fernhill Travel in April 2005, but left in September of that year, running in this livery. Note the offside door, which functioned as an emergency entrance and 'continental' door if abroad. Coaches had got generally of a higher build, whilst wing mirrors also got larger. W808 AAY is seen in North Street yard.

Top – Another local coach firm with a 3-generation history is Hodge's Coaches of Sandhurst. From them came another new make to take the place of Bedfords, the German-built MAN 18.370-type. That firm had started production in 1890 and at one time had Rudolf Diesel on the staff, whilst the Dutch-built Berkhof 'Excellence' 1000L 53-seater body was another first at Winkfield. New in 1994 it came in June 2003, and is seen that same month still in the dark blue and bronze livery of its former owners and on Route 83 for Cox Green School.

Middle – With White Bus the coach seen above and here after full repainting ran as L522 MDP, but with Hodges it was 1598 PH, one of their 'cherished' collection of registrations, the PH representing Peter Hodge, the head of the firm at the time. Similar coach L715 FPE, which is shown elsewhere, had re-gained its original Surrey mark after running as 5881 PH. L522 MDP clearly shows the offside additional door and the access to luggage lockers.

Bottom – The search for suitable types for service bus use continued, and in September 2007 several demonstrators were tried out, including this one seen emerging from Rangers Gate on Route 01. Dennis of Guildford had re-entered the PSV market, and in collaboration with Alexander coachbuilders had created the Enviro 300 of which SK07 DYA was a 44-seater version. White Bus had previously had a long association with buses of Dennis manufacture.

Top – Dennis chassis did indeed make a return to the fleet, though in the form of coaches. A quartet of the Javelin-type was acquired during 2005, and further illustrations feature in the black-and-white photos. R674 OEB carried a Berkhof 'Excellence' body which seated 57 and was new in 1997. The Javelin featured an under-floor Cummins 6-cylinder engine of 8.3 litres and was aimed at the former Bedford and Ford PSV market. 'OEB' was used on contracts and private hire, where its high capacity was useful.

Middle – Here we have the opportunity again to see the inside story of how coach bodies developed over the years, with a view of 'OEB' towards the rear. The seats are comfortable but practical in order to achieve the higher number, whilst the lack of opening roof vents eliminates the likelihood of leaks and un-wanted insects. When it is considered that this photo was taken when 14 years old, the good order of the interior is a credit to White Bus and its maintenance staff.

Bottom – A further two Dennis Javelins were added in 2006, and here we see L321 XTC on the right at North Street Garage. With a Wadham Stringer 'Vanguard' MkII body, it had been new to the Military, and quite a bit of work went into turning these into 70-seater buses for school's contracts to meet increased seating needs on some services. Note the permanent school bus signs and lack of indicator for route on this type. Next to it is Bedford YNV-type E849 AAO.

Top - *The other Javelin to arrive in 2006 was from a civilian source and carried a Berkhof 'Radial' 53-seat coach body. R714 KGK was new in April 1998 and introduced yet another body and chassis variant. It is seen at Reading Road Pond on the way to the Cobham Bus Rally in April 2007. The use of school bus signs shows how it was mainly used, and these were specified in contracts by Local Authorities, whilst the Highway Code requires drivers to exercise extra care when passing such vehicles.*

Middle – *Of the other ex-military Dennis Javelins, L606 ASU is seen at North Street. Their lack of route destination screens was not a problem, as route boards were used on contracts and registered school services. The 70-seat capacity was very useful for in-between hires by schools to local places of interests and the London museums. Buses of this type also came onto the 39 from Finchampstead to St. Crispins School in Wokingham, as capacity had previously required two vehicles.*

Bottom – *We have already seen that generally vehicles were referred to by their registration marks in the modern era, but in the re-registration for civilian use two of the Javelins became 'ASU', the other being L561 ASU. This October 2005 view shows the length of these vehicles, whilst the bodies also featured a pair of rear doors, which in army use could take stretcher cases. At the time White Bus acquired them the issue of accessibility for students in wheelchairs etc. was a hot topic.*

Top – *Optare developed a range of integrally-built buses, including this Solo M880-type, which saw widespread use in both town and the countryside. Although White Bus would indeed order some in due course, YJ57 YCF was in fact on loan in place of YJ07 EGD, which was away in May 2008 for some warranty work. It had a 22-seater body and leaves Cranbourne Gate with 'Town Centre' in place of the more usual Windsor when covering Route 01 from 27th-31st.*

Middle – *An Optare Tempo X1260 demonstrator was tried out in November 2007 and led to two of that type being purchased. When Route 01 was re-launched in January 2009 they were repainted as seen here on YJ57 EHV outside the new stand at Ascot Race Course on a publicity run. The old Borough Bus signage has given way to the RBWM crest, seen next to the White Bus Services on the sides, though the wording is retained on the front panel.*

Bottom – *The new scheme incorporated the names of the areas served by the route, whilst the designers had produced a scheme which incorporated local points of interest served. Unfortunately though the 'contra-vision' application of vinyls over the windows made seeing such places difficult, whilst the author makes no secret of his dislike for such schemes, believing that the view from the window is the best (especially in the Park), and also quite useful for knowing where to get off! The location of 'EHV' is The Village post office.*

Top – *Mention has already been made of the Alexander-Dennis Enviro 300, and one of those seen had been the 60-seater school bus version back in August 2007. Several such examples had been taken by Courtney Coaches of Bracknell in September 2005, but changes to some of its school routes led to them being ousted by double-deckers. Two came to White Bus, the second of which was SN55 DVA, and it came in May 2009. The 'Cool Bus' livery had been developed by the makers to appeal to children, whilst the yellow base mimicked the iconic US-type vehicles also seeing some use then by First Berkshire locally.*

Middle – *The other of the former Courtney school buses was SN55 DVB, which arrived in April 2008 and is seen in the North Street yard. With 60 seats Courtney's had often faced complaints regarding standees from parents on the Holt School route, so replaced them with 76-seater double-deckers. The subject of 'yellow' buses was another hot topic in home-to-school transport at the time, with First offering some genuine US buses converted to right-hand drive, whilst other manufacturers offered new models for that purpose.*

Bottom – *White Bus was also able to add a further example from the in-house fleet of West Sussex CC in October 2011 in the shape of MX09 GZL, which carried this plain yellow and black livery. It is seen at the Hawthorn Hill out-station in March 2013, the yard space at North Street having reached saturation with an expanding fleet.*

127

Top – Whilst awaiting new buses for the additional routes in Windsor, it was necessary to hire two buses from Dawson Rentals from February 2014. Both were on Hungarian-built chassis by Enterprise Bus and carried Plaxton 'Primo' bus bodies. One was KX57 FKW, seen here at Datchet on the extension of Route P1 park-and-ride, which got off to a shaky start with extensive flooding in the Datchet area. Both buses left in March when those intended for the route were delivered.

Middle – A pair of Optare 'Solo' M790SE-types was ordered for the Windsor Local Services, with 27-seater bodies, and they were received in March 2014. YJ14 BBN is seen here by The Green at Datchet on one of the extensions of the P1, with Krippasur Rai taking his short break. This rear-engined design makes good use of space, with the usual modern-day interior mix of fixed and tipping seats, spaces for buggies and wheelchairs and CCTV throughout.

Bottom – The other Spring arrivals for 2014 were a pair of Optare 'Versa' V1170-types, fitted with 44-seater bus bodies and received during April. They took over the duties on Route 01 and YJ14 BBV is seen leaving the Great Park through Ranger's Gate onto the A332 for Windsor. These buses conform to the access requirements of the LA's in support of the route, whilst note the small wheels and long wheelbase of these 11.7-metre buses. The 2014 deliveries all wear this revised form of livery.

The year 1994 was notable for the first cover for Mark Way of *Reading & Wokingham Coaches,* along with others for Slough-based *Lianne Coaches.*

On the schools front St. George's of Ascot had in total 105 hires, with destinations as diverse as the London theatres and museums, the Newbury Show, River Thames boat trips, the Ideal Home Exhibition, Coral Reef Water Park at Bracknell, The Guard's Polo Club and shopping at Knightsbridge. Younger parties took in the new centre at California Country Park, Wylde Court with its Rainforest near Hampstead Norris and the Chiltern Open Air Museum, whilst adult themes included Brooklands Museum, the Cabinet War Rooms at Whitehall, Sandringham and the Redgrave Theatre at Farnham, there being over 200 different destinations in the private hire diary for 1994!

It was also the first Summer of an exceptional project for foreign students run by Nord Anglia Education, in transport terms requiring 4 or 5 full-size coaches each day during July and August. The students were based at Kingswood Hall in Egham, and the task of supplying the transport was given to Doug Jeatt, who elicited pledges from *Fargo, Fernhill, Tent-Tours* and *Windsorian* as he charted out the daily requirements, which took students to London, Oxford, Winchester, Stonehenge, Brighton, as well as for leisure at Thorpe Park and Chessington World of Adventure.

Another unusual job saw *White Bus* providing a direct link between Reading Station and Farnborough Air Show in September 1994 on hire to the Railway and operating a shuttle from 07.30 through to 18.15.

Incoming in February 1995 was this Bedford YNT with Plaxton 'Paramount' MkII 53-seater coach body (B633 DDW), which originated in Cwmbran but came to White Bus after service with Fernhill Travel. Here we see it passing another local public house, now as flats, the Fleur-de-Lis, one of many in the area with a Royal connotation. It is also worth noting that it was one swapped by Courage with Morland by then.

1995 was the year in which the old 'Setright' Ticket Machines would be replaced by the first of a new generation of 'Wayfarer' machines, which also allowed better analysis of passenger journeys.

The daily transfers for HFC Bank ceased in early March as a result of the end of an agreed period after the relocation but, as happens, another schooldays run for LVS started from 13th of that month. This was for pupils whose parents paid for transport, registered by *White Bus* until September 2001, when the school put it on a contracted basis. Known as Route 09, it started at Camberley and ran to the school by way of The Maultway – Lightwater – Windlesham – Sunningdale (Station) – Broomhall Lane – The Rise – Sunninghill (High Street) - Berystede Crossroads – South Ascot – Ascot (Station) – Heatherwood Hospital. As can be seen, it now covered the morning 07 link between Ascot Station and the school, whilst in due course the route was further back-extended to take in Crowthorne, Sandhurst and College Town.

The first of another regular annual job took place in May 1995, when a party from Belgium were met off the Ramsgate ferry for the Nematological Conference, an event for specialists in the study of Roundworms, and the party visited various relevant sites at Silwood Park, Kew, Rothampstead and St. Albans during their 5-day stay, which ended with time in Canterbury on the way back for the ferry.

Transport in connection with the Guard's Polo Club at Smith's Lawn had involved *White Bus* in various ways over the years, and no doubt that led to the hire by the Venezuelan Embassy Polo Club for a trip down to the ground at Cowdray Park, near Midhurst in West Sussex.

Over at Bracknell *Fernhill* had got a new job with a shuttle service between Bracknell Station and the Mercury Communications building off Western Road, a task quite often covered on their behalf. Of course such cover often brought direct enquiries, and after a number of runs for other operators from the Abbey School in Reading, a booking followed to take a party of girls to Heathrow for a Munich flight in February.

On the staff front Ray Shears joined in what was a busy year for private hire. Imperial College had long used the firm for transfers between Silwood Park and other venues, but during 1995 a coach went several times to Pulborough Wild Brooks, where the flora was surveyed as part of the process of it becoming an RSPB reserve.

Regular events organised by the Berkshire Youth Music Trust were also catered for, which also needed some allowance for the instruments to be carried! In other cases, the need for specific vehicles arose for such things as a microphone or a wider entrance door.

From September yet another vehicle was needed on Charters Route 24, so a 24B variant was forthcoming, the actual stops covered by that and other variants changing over the years according to assignments.

There was addition to the fleet from October 1995, when another Bedford YNT-type arrived, but this time with a bus body (C668 WRT). This 1986 example had a Duple 'Dominant' body, though in its case with some 3+2 seating towards the rear, giving a very useful capacity of 63. It was acquired direct from the first owner, *Chambers of Bure,* in whose dark red livery it was initially pressed into service.

'WRT' was another good investment, remaining in service for 12 more years. When repainted it was also the first to wear the new scheme which featured a pale grey skirt, as seen in these views. <u>Above</u> we see it passing under the Goswell Arch, where the Windsor – Slough line leaves the Central Station on its viaduct, and again at Woodside, by the Rose & Crown <u>below</u>.

The year ended with the annual Christmas Service for the girls of St. George's Ascot at the Parish Church in Windsor, which called for 7 coaches, 3 of which were hired in from Windsor-based *Tent-Tours.*

There was an increase in rail-replacement work during 1996, when various sections from Maidenhead on the branch to Bourne End or on to Marlow were upgraded usually at weekends, though sometimes also after the evening rush-hour. *White Bus* was suggested by Sean McAleer, who worked on the trains, but also later on came to do some work for the firm. There was also cover on the Windsor – Slough branch at Christmas.

As a result of the arrival of 'WRT' in late 1995, YRQ-type Willowbrook-bodied bus 'GNV' departed in February 1996, going to a Children's Entertainer of whom we shall hear more in due course.

What had been the Charters Route 24B was given the new designation 38 (or M38 at RBWM) from 29th April 1996, though further variants of the 24 would appear later. At that time the 38 had only two pick-up points at Cranbourne Cottages and The Cranbourne Tower pub, both in North Ascot, with 45 assignments.

In the meantime *H&H Coaches* had sometimes needed cover on its Cox Green School run from the Holyport area, which it had replaced *MD Coaches* on a little earlier. That was from March 1996 onwards, and Ian Harris duly had to give up the business which resulted in *White Bus* gaining the contract from Autumn Term on 23rd September. However, it seems that Doug never got paid off for all his covers, so Ian Harris worked for him for a short time (before going to Norfolk it is said), plus his pink-painted Bedford YMT-type with Duple 'Dominant' 53-seater coach body (ODL 632R) came into the fleet in September, though it was only kept until December of that year. A photo of this coach appears on page 94, and more private hire followed from Cox Green School.

Whilst mentioning school contracts, another aspect to consider is how routines could be upset by variations in closing times at term ends, the general practice for Berkshire schools being to finish around mid-day, with the difficulties that caused when more than one establishment was served by a route. At the other end of terms, some schools favoured staggered starts for various year groups, a logistical nightmare for both the operator a local transport co-ordinator.

For some four years the former Safari Park had been closed as it was re-developed into the UK Legoland similar to that long established in Denmark. When it was opened in March 1996 a quartet of 53-seaters was used to bring guests from Windsor to the site. In a generally busy year for hires St. George's had 105, LVS had 68 and Charters 64, with the former on an extended trip to Slapton Sands in June. A party from the Oxford Forestry Institute toured the Great Park, whilst school groups visited the Transport & Road Research Laboratory, Crowthorne.

We have already heard about the re-development of some of Windsor Town Centre, but in reality the options were quite limited when Tesco Stores wanted a presence in that town, with space severely restricted by the Castle, the Great Park and the River Thames. It therefore settled on a site to the west of the town in the Dedworth area, after which some passengers approached *White Bus* about providing a link. Although many people did by then have cars the 'big shop' was still the norm for most working people.

In response to this request *White Bus* made a Fridays-only extension to the Route 01 journey which arrived in from Ascot to the Theatre Royal at 10.18, whereby it ran on under the Goswell Arch and the Maidenhead Road to the new store. It then left at 11.45 to form the 11.55 through to Ascot, with the service starting from 3rd January 1996, charging a return-only fare of £1.75.

'KDT' waits at the Dedworth Tesco Store, though the sole 'passenger' was another White Bus driver, and it will be no surprise that after a short time it ceased.

Another coach came in March 1997 in the shape of a 1985 Bedford YNT-type carrying a 53-seat Duple 'Laser' body (B542 OJF), after a number of owners.

There was also a smaller addition to the fold in July, with an Iveco 43.10-type, which had a 'City' 21-seater body by Robin Hood (D473 WPM), a former member of the *Alder Valley* and *Stagecoach South* fleets. Prior to that any smaller party hires were given to *J&J Coles* of Yateley or *Fernhill Travel,* and in fact Frank Holgate from the latter also hired 'WPM' at times to cover their contracts to Presentation College on the Bath Road west of Reading, or to Kennel Lane School in Preistwood. It was also used on some of the lighter service runs and the 02/03 school journeys.

In respect of work undertaken by *White Bus* for other concerns, there was a 2-coach hire from Mychett to Odiham on behalf of *Countrywide Travel,* whilst rail replacement saw 3 buses used on Sundays to cover the Windsor to Slough branch a number of weekends. The Nord Anglia Summer School called for 61 coaches over 27 days during July and August, with 5 vehicles needed some days, and assistance came from *Fernhill* and *Reading & Wokingham Coaches.*

Another wedding saw two coach-full of guests ferried between the Royal Berkshire Hotel just east of Ascot to St. Peter's Church in Hatchet Lane, Cranbourne and to the reception during July, whilst Bob Napper continued to organise outings from the Alpha Arms in Slough to Maidenhead FC away matches and to the Newbury Races, as well as those for The Garibaldi at Chalvey and the Working Man's Club, driven by him. Indeed, the whole age range was represented, from the Garth Under-5's to the Eton Wick Over-60's, whilst Whitekights School at Earley booked for a 6-day study break to Swanage in April/May 1997. Schools also hired for airport transfers, with Cox Green going to Montreal, The Abbey to Dusseldorf, and pupils of the LVS flying down to Johannesburg!

Two of the 1997 acquisitions are shown in this view of the North Street yard. On the left we have Bedford YNT (B542 OJF), still in a livery of white with stripes in medium blue. The little Iveco (D473 WPM) on the right has now been repainted into the white and grey scheme, whilst in the background stands 'SNK'.

More locally the Waltham St. Lawrence School hired for a tour of the Great Park, as did St. Peter's of Old Windsor and St. Francis RC School of South Ascot, whilst topical venues for 1997 were Clink Street in the refurbished Docklands of Southwark for the London Dungeon, Cadbury's at Bournville near Birmingham and the Polka Dot Theatre in Wimbledon, the latter with 3x53's for Marist Juniors. Some hires also came from St. Bernard's Convent School in Slough after a number of covers for *Alf's Coaches* earlier to there.

On the adult hires, there were newer theatres such as The Peacock in Woking and the Corn Exchange over at Newbury, whilst the Burma Star Association went from Slough to various places, and the Locomotive Society from Ascot to see the beam engines at Crofton on the Kennet & Avon Canal west of Great Bedwyn.

On the contracts front, a new one came about through the merger of Haileybury School, which had for many years been at Clewer Manor, off Imperial Road in Windsor, with Lambrook School at Winkfield Row. Transport was provided for those students transferred from 10th September 1997, and it inevitably led to more hires from the new enlarged Lambrook School.

Also with school work in mind, the coaches of the fleet started to be fitted with seat belts, with 'KDT' so treated in October 1997, though still not required by law at that point in time.

During that year Les Birchmore, Ian Dundas and Allen Titchenor joined the staff, but on the vehicle front no disposals took place in response to the additional contracts now being covered.

From time to time vehicles were borrowed from other operators, and on 15th September 1997 Tony Wright recorded the presence of a blue-liveried Bedford YMT at the *White Bus* premises in North Street. Although he took a number of photos of it with Joe Sutcliffe at the wheel, no one can now recall why it was present.

The mystery blue coach (YKO 653S) is seen outside the Hernes Oak, at that time owned by Shearer of Maybury and bearing the legend 'Earl of Surrey Battle of Bosworth 1485', though actually built 1978.

Acquired in December was 1983 Bedford YNT-type (JNM 747Y) with 53-seater Plaxton 'Paramount' MkI coach body, which came from Grey's Coaches of Ely, in whose cream and green livery it was at first used.

Another vehicle loaned was in September 1998, when Edward Thomas & Son of West Ewell provided HIL 4017, a Plaxton-bodied Leyland 'Tiger'. It is seen working the school Route 13, and is in the process of also exchanging children with the Iveco bus 'WPM' at The Village Post Office during the afternoon journey of Route 03. Such a connection maintained a greater variety of options for school destinations without wasted mileage.

1998 saw an increased use of little 'WPM' on smaller hire jobs and loans to *Fernhill*, whilst *White Bus* also still deputised for that firm on the shuttle between Bracknell Station and Western Road on the 'Mercury' contract, though that company had now passed to Cable & Wireless.

Another contract to Charters School started from 19th January 1998, known as Route 40. From September the 24 became 24B and the 38 became the 24A. All of these had 'M' prefixes according to RBWM, though *White Bus* only sometimes recorded them as such.

On some very localised hires for schools, the transfers were arranged as relays, and two examples from 1998 are Papplewick School from Windsor Road by Ascot Race Course down to Blacknest Gate, and for the carol service from Lambrook School to Winkfield St. Mary's Church, each journey only around 10 minutes. Another Christmas service that year saw 360 pupils from St. Edward's & Royal Free taken to the Parish Church from Parsonage Lane, again over several runs.

School journeys of longer duration in 1998 saw Lea Junior of Slough off to Overstrand in north Norfolk for a week, though the coach came back in between. A party of girls from St. George's went off to Slapton Sands in South Devon but, unfortunately, 'DDW' had a problem on the M5 near Taunton and a replacement had to be organised from a local operator. The coach was towed back by Warfield Garage, and it was found that a woodruff key in the drive to the fuel injector pump had sheared, the vehicle returning to service the following day.

In respect of the fleet, April 1998 saw the departure of YRQ-types 'SNK' and 'STL', which had clocked up 24 and 17 years respectively for the Company. The former will of course be recalled as the bus ordered when Doug left Terry and Richard looking after the place, whilst the latter came secondhand.

On the staff front Phil Bathard arrived in April 1998 as part-time, and Ashley Titchenor in July, whilst Helen Taylor came as Book Keeper in October. It is worth noting that Phil, as well as Brian Cox, Peter Ives and Bob Napper had all worked at Heathrow at some point as drivers for various companies.

Mention has been made of the extensive hires by St. George's School of Ascot, but it should be noted that as a boarding establishment, many were also run at weekends, with leisure activities at Coral Reef and the John Nike Centre, whilst in term time a regular event saw girls taken to Pangbourne College for the evening, with a return pick up at 11pm. Other notable ventures from the school included Wimbledon for the tennis and a trip on the tethered balloon at Battersea, which offered views over London before the Eye.

Over at Wokingham the all-weather pitches proved popular at Cantley Park, particularly as fixtures often had been cancelled due to the poor condition of grass pitches, which was also a nuisance to the operator. Cox Green School took a trip to the Arsenal Soccer Academy, whilst other school parties tried the Shire Horse Centre near Littlewick Green and the RAF Museum at Hendon in North London.

Rugby-related hires saw the Grenadier Guards going to London Welsh and Bracknell RC to both Wasps and to Twickenham. Mark Timms of Ascot Golf Club organised several trips each year to Alicante, hiring a coach for the transfers to Gatwick or Luton, whilst on the Catholic grapevine a party from St. Margaret Clitheroe RC Church in Hanworth, Bracknell used the firm for its journey to Heathrow for a flight to Rome. The Polish Community in Slough went for a coach tour of Henley and Stonor, the Chavey Down OAP's went to Poole for a week, and the Winkfield Bell-ringers toured 8 churches to end up in Bognor Regis. Another wedding involved a journey picking up at St. Crispin's in Wokingham for Weybridge.

The only vehicle added in 1998 was another Bedford YNT-type, though fitted with a Duple 'Laser' MkII 53-seater body (B30 MSF), which had previously been used on luxury tour work in Scotland by Silver Coach Line of Edinburgh, but came to White Bus via others in May 1998 when 14 years old. This vehicle had quite a lot of work done on it during the Summer months which included the fitting of a Cummins C-series engine, recalled as a tight fit! It is seen on the South Devon coast in June 1999 with Mick Fazey on the St. George's School Slapton Sands break.

On the subject of Mick and his activities, a time sheet gives an interesting insight to the way in which driving and other duties combined to good effect when keeping the jobs covered and the fleet in trim.

'Wednesday 6th July 1999: 7.15 - 9.00 Charters 24A and 24B; 9.00 – 13.00 in yard fixing door rubbers and others to driver's window, lubricate catches to engine flap and clean racks, all on 'JMJ'; 30-minute lunch break, Licensed Victuallers School to Bracknell Sports Centre and John Nike, then onto St Crispin's School Route 39 afternoon run; pick up LVS and take back to school; late run for Charters School, finished at 17.30, total hours worked 9 hours 45 minutes'. The rest of that week shows how he was bringing 'JMJ' up to scratch whenever not engaged on driving duties.

As noted early on, the White Bus involvement with the County school swimming sessions goes back to pre-war days, and that continued to evolve in terms of the schools served and the venues used. By 1999 indoor pools at Langley and Montem Leisure Centres, to the east and west of Slough at were in use, whilst the focus had shifted to the schools around that town, with the following Primaries over the weekly cycle –
Baylis Court, Castleview, Cippenham, Colnbrook, Farnham Royal, Godolphin, James Elliman, Montem, Our Lady Of Peace, Priory, Ryvers, St. Anthony's, St.

Ethelbert's, St. Mary's, Slough & Eton, and Wexham. Although the list looks lengthy, it should be appreciated that many of these schools were close by each other, so changeovers and mileage was minimal.

Mention of Wexham also prompts the comment that the Hospital there had an active Theatre Group, which travelled to London venues regularly, the coach also picking up at Heatherwood Hospital, for staff that is! The Lin Carpenter School of Dancing took 2 coaches full up to The Haymarket Theatre in London, whereas Ascot United FC had a trio to Crystal Palace, both in March 1999.

The contract for Cox Green School M03A was registered by White Bus with effect from April 1999, and was given the vacant number 83, which was appropriate to bus services in the Maidenhead area. A number of covers for operators that year included the Courtney contract from Maidenhead to Jealotts Hill for ICI workers, as well as for Frimley Coaches.

Over in the Wokingham area the number of pupils on Route W39 to St. Crispin's School travelling by the end of the Summer term could at times exceed the seating capacity. The situation was complicated by the route serving an area which straddled the distance laid down under the Education Act 1948, whereby pupils under 3 miles safe walking distance from the school were not funded, so parents bought passes direct from White Bus, which had registered the service from 17th February 1997 at the suggestion of Brian Coney, who at that time was re-organising the supported bus routes in the Wokingham DC area. That move also secured tenure of that service for White Bus through to the present day.

The route was also interesting as it started at Luckley Road on the Finchampstead Road, on the south side of Wokingham, and then ran away from the town to take

in Handpost Corner – Pine Drive – California Cross-roads – Kiln Ride – Kingsmere Lake – Nine Mile Ride – St. Sebastian's Hall, after which it ran through Gardeners Green and over the Waterloo Crossing on the Reading to Waterloo railway, before turning down Rances Lane to emerge at the school on London Road. On the homeward journey the vehicle left the school towards Wokingham town centre, along Peach Street and down Denmark Street onto Finchampstead Road and under the two railway bridges, from where it took the same route to St. Sebastian's as in the morning.

From 4th October 1999 the route had two vehicles and was split as W39A and W39B. Indeed, the situation was becoming acute once other local bus services saw further reductions, making it the only viable link.

No vehicles left the fleet during 1999, but two coaches arrived, which reflected the increased activity. In October a 12 year-old Bedford YNV-type with a Duple 320-style 53-seater body (D259 FRW) came via a number of owners. It was joined in November by another YNV-type, but with Plaxton 'Paramount' MkIII 53-seater bodywork (E849 AAO) of late 1987, and which started life in Carlisle, but also via others.

'FRW' still in red and white livery at North Street.

Above we see 'HRO' still going strong on the service Route 01, as it travels along Kings Road and parallel to The Long Walk on its way out of Windsor, the Round Tower being the most prominent part of the castle. From the 6th floor of a building in Reading the tower can be seen on a clear day, a reminder of how extensive the original Royal Forest once was!

On the subject of the Royal County of Berkshire, it had ceased to be the Local Authority, being replaced from 1st April 1998 by the elevation of the existing 6 Borough or District Councils to Unitary Authorities, though only Newbury actually altered its name to West Berkshire. From the point of view of *White Bus*, there was little immediate change as the daily work on transport had been out-sourced to a firm called BABTIE back in 1993, a less than satisfactory arrangement. From 2000 the individual UA's set up transport sections with their own staff, some of whom were former BCC. All, however, were now dealing with supported public transport, home-to-school and special needs transport, as well as promoting public transport through area guides and maps.

At the close of 1999 the fleet consisted of 13 vehicles and, in order of date new were: 1975 - KDT 281P; 1977 – RLW 778R; 1979 – HRO 958V and EAJ 327V; 1980 – JMJ 633V; 1983 – JNM 747Y; 1985 – B542 OJF and B30 MSF; 1986 – C668 WRT and D473 WPM; and 1987 – D259 FRW and E849 AAO. All of the above were on Bedford Y-series chassis, other than the 21-seater Iveco (D473 WPM), whilst only 'HRO' and 'WRT' carried service saloon bodies.

Ray Sutherland joined as a driver in 1999, and a contemporary article stated that at that point there were some 20 drivers employed, only 7 of which were full-timers. Flexibility in the varied coverage was maintained by keeping in touch with those who could be called upon when extra drivers were required.

Another vehicle still soldiering on was 'KDT', then 24 years old and acquired with Winkfield Coaches. It is seen passing The Duke of Edinburgh at Woodside and was kept in good order. The Plaxton 'Panorama Elite' MkIII body only seated 45, so it was often used for smaller private hire jobs. It was another vehicle still with some life in it when sold in favour of newer types, so it is pleasing that it made it into preservation.

2000-3 – European Influences

The millennium year saw the coaches of *White Bus* as busy as ever, and in February 4 were needed to take the staff of the Crown Estates Office for an evening at Ascot Race Course. The Women's World Day of Prayer saw a trio of vehicles arriving at the Royal Chapel from various points, whilst other parties went to The Dome at Greenwich. Both the Royal East Berkshire Agricultural Association and the Country Landowner's Association undertook tours of the Great Park, whilst the latter also went on to see the new lake excavated on Eton College land at Dorney, which had the dual purpose of providing a flood-relief for the area, whilst also being used by the college as an alternative to rowing on the busy River Thames, plus of course it would see use in 2012 for the Olympics.

The Denbies Wine Estate near Dorking was a popular new venue with adult groups, as were the Wimbledon and Oxford Greyhound-racing circuits. The LVS had a trip to paint-balling, whilst amongst the pre-school groups such names as Tiny Boppers and Little Fingers featured amongst the hires.

No vehicles entered the fleet during 2000, but several departed, with 1977 Bedford YMT-type (RLW 778R) leaving in March, followed by the 21-seater Iveco of 1986 (D473 WPM) in April, although the latter was not replaced by something of similar size.

Unfortunately, April also saw the passing of Mrs. V.J. Jeatt, who died on the 18th of that month at the age of 84. Although not directly involved with the business, we have already seen that without her support the bus service would no doubt have ceased many years previously.

On 19th May 2000 Alan Moore retired, with an evening 'do' at the Hernes Oak, and his place was taken by Les Copley, after which he, along with Greg Edwards and Geoff Lovejoy were those usually found on the service route. Greg had recently arrived, and in that year they were joined by Brian Lewis and Martin Shaw, the latter at that point being a driver.

First Beeline reduced its services in the Sunningdale area from October 2000, leaving only a pair of timings running beyond Ascot, plus each way to Charters and now as part of a revised Route 192 (Reading – Earley – Wokingham – Bracknell – Warfield – Brookside – Ascot – Sunninghill – Sunningdale).

Outgoing in 2000 was 'RLW', which had come through Winkfield Coaches a decade previously and proven itself as another Bedford workhorse. Note also how much bright-work adorned the Duple coachwork.

Bedford YNT-type saloon (C668 WRT) was caught as it turned out of Church Road in Chavey Down onto Long Hill Road, displaying Sunningdale 07 on the blind, though that was not actually part of thet route.

The longstanding school Route 02 (St. Peter's School, Old Windsor – Windsor – Parsonage Lane, Dedworth) and 03 (North Street, Winkfield – The Village, Great Park – Royal Lodge – St. Peter's School, Old Windsor) were no longer required in that form after the end of the Spring Term on 6th April 2001. However, Route 13 continued to serve the points in the Park for schools in Windsor. On the Charters contracts, there was an additional 24C variant added from Summer Term on 23rd April, plus from the next term a 24D was also required as numbers increased.

Also during that same month one of the many events held in that town affected traffic and therefore buses at times, that being the Colours Ceremony, with troops marching from Victoria Barracks to the Castle.

Among notable hires for 2001 were some regular ones taking parties from the Thames Hotel in Windsor over to Legoland, whilst the new arts and performance site at Norden Farm, just to the west of Maidenhead, had a large-scale visit from Charters School on 4th January with 4x53-seaters needed. The Milestones Museum at Basingstoke also now featured for both adult parties and school trips, as it re-created old street scenes and celebrated local industries. Older people also saw an increase in daytime clubs, some examples making hires being The Monday Club and the Tea@3 Club.

During February 2001 the Chris Whitley Band hired a coach for a gig in Islington, following that up in April for another at The Swan Theatre in High Wycombe. There was also another wedding hire, with 2 coaches collecting guests from Maidenhead and Hurley, then to the church at Cookham, and afterwards to the New Mill at Eversley for the reception.

More tours of the Great Park were booked, with the Campaign for the Preservation of Rural England there

in June. The White Bus Services Enthusiast's Group also attended a particularly relevant event, the Bedford Gathering at Cambridge on Sunday August 26th, by which time the fleet was again 100% Bedford!

The tally of schools having swimming transport had also increased during 2001, with Western House (Cippenham), St. Nicholas (Taplow) and Wraysbury School all going to The Montem Centre in Slough.

On the staff front Tony Davidson arrived in February 2001, followed by Eric Smith in October.

Only one vehicle entered the fleet in 2001, when in August what was to be the final Bedford acquired was purchased from *Fernhill Travel.* It was also the first to bear a foreign-built coach body, being a 1986 YNV-type carrying a 53-seater Belgian-built Van Hool body of the 'Alizee' style. Like a number of other coaches of that era it had acquired an 'Irish' mark (TIW 2795), a common ploy for disguising the year of origin.

The former Fernhill YNV (TIW 2795) ran for a time in their livery of deep red and white before receiving a re-paint into white with a grey skirt panel.

There were no changes within the fleet during 2002, but in the early months Brian Lovejoy ceased his role on maintenance, so Andy Canning was taken on as Fleet Engineer from March. Other full-time drivers to arrive that year were Frank Hamilton and Mike Hoare, whilst Carol Masters and Michael Poynter did some turns but neither for long.

The first trip is noted to the London Eye during 2002, whilst Jagz at The Station (a music venue/wine bar in the former Station Hotel) in Ascot booked some trips to Wimbledon Dogs. The Aerial Motorcycle Club also favoured *White Bus* for its outing to Brooklands and its collection of historic motorcycles, and in Bracknell the U3A made regular bookings for day trips to places of interest and theatres. Work done for Meadowbrook Montessori School evidently led to the Montessori at Littlewick Green also hiring.

Covers included those for Ian Hughes *(IJH Coaches)* on his Charters Schools contract, as he tried to maintain his small operation with some difficulty.

This page is dedicated to 1986 Bedford YNT-type as it would have the honour of being the final saloon bus of that make in the fleet of White Bus Services.

Top –Seen at the bus stand opposite the Theatre Royal in Windsor, the 63-seater Duple 'Dominant'-bodied bus was working on Route 01 to Ascot on a journey via the Royal Lodge. Also behind is Willowbrook bus-bodied 'STL', which left back in 1998 after 17 years in the fleet.

Middle – On this occasion we see the bus emerging from Lovel Lane by The Crispin after one of the journeys then operated via Fernhill. The area hosted a number of hamlets which developed outside the area of Windsor Great Park, whilst The Crispin is now very close to the peanut-shaped roundabout which divides the roads taking traffic to Winkfield and Legoland and that passing through the Great Park directly to Windsor. It was an experiment by the people of the Transport & Road Research Laboratory over at Crowthorne, one of many undertaken locally.

Bottom – This nice example of a location shot, complete with street sign, tells us this is Sunningdale, which was close by the railway station. Here we see Geoff Lovejoy coming back to the bus with his daughter. The photo also shows the front entrance, with steps up to saloon level, and also the board with route details carried on the nearside. This bus would remain in service until December 2007 after 11 years service with the Company.

The steady increase in the fleet led to the need by late 2002 to outstation some vehicles, so the Lorry Park in Bracknell off Longshot Lane became used for at first a pair of contract vehicles, joined in due course by a third. Tony Wright didn't take many views there, so I am grateful to Michael Clancy for this shot of 'WRT', 'JMJ' and 'DDW' allocated there for such duties.

On the school contracts, *White Bus* added another from 12th June 2002, as offered by Eric Mouser of RBWM after Trevelyan School was relocated to the former Royal Free site, and it was registered by the Company from the outset of its cover as Route 88. It provided a link from various points west of Windsor to a number of stops in that town, starting at Fifield (Meadow Way), via Oakley Green (The Green Oak) – West Windsor (Ruddlesway) – Testwood Road – Gallys Road – Longmead – Dedworth (Smiths Lane) – The Maypole – Bell View – Clewer (The Three Elms, for St. Edward's School) – Orchard Avenue (for Windsor Boy's) – Windsor (St. Leonards Road for Trevelyan) – Imperial Road (for Windsor Girl's). Also starting in September 2002 LVS Route 09 from Crowthorne and ran via Sandhurst if required.

Geoff Lovejoy and his wife decided to move to the West Country, so on 5th July 2002 he left *White Bus,* though they subsequently returned to the area and he resumed his duties from 4th August 2003.

Some rather exceptional hires came during March 2003, when the Royal Chapel Choir went to the Royal Albert Hall on the 1st, then 2 days later parties from Charters, Windsor Girl's and St. George's Ascot went to Reading University, the latter requiring 3 coaches of *White Bus,* plus a trio from *Fernhill* and one from *IJH Coaches.* On a later date to the Royal Albert Hall, the whole of the Royal School decanted there in 4x53-seaters, with one coming from *Fernhill* and *IJH.*

It should be appreciated that production of Bedford PSV chassis ceased in 1986, and whilst *White Bus* had plenty of spares, no low-floor models had been built.

Therefore, when a new generation of saloons was required for the 01 route, RBWM called for low-floor buses to be provided from April 2003. The choice was the Optare 'Delta', a collaboration between Dutch-based chassis builder DAF and the Leeds-based bus bodybuilder, as both firms were then in the same group. The SB220 chassis had an Alusuisse integral-type body built on to achieve the lower floor level. The power was provided by the DAF 11.6-litre 6-cylinder diesel engine, with power steering and ZF automatic transmission. These buses would prove popular with the drivers, as well as to the management as they returned 8.5mpg in service, whilst their 49-seat capacity and easier access was useful all round.

However, despite the above accolades, those purchased were both built in 1998, with one new in April (R89 GNW) and the other in September (S158 JUA). Each had previous lives on airport-related work, with 'GNW' being used at Dublin Airport for *Flyer Bus,* and came back through Optare, hence the Leeds area mark. 'JUA' was also a mark from the same source, but came from nearer home, having been based at West Drayton for the *Speedlink* air-side operations at Heathrow.

Optare 'Delta' R89 GNW is seen later on in Windsor.

139

Yet another new make followed in June 2003, when a nice pair of MAN 18.370-type coaches was purchased direct from *Hodges Coaches* of Sandhurst. New in April 1994 they carried Dutch-built Berkhof 53-seater 'Excellence 1000L' front-entrance bodies on German-built chassis. Peter Hodge had been acquiring marks with 'PH' in them and putting them on the vehicles for some years, so these had run as 1598 PH and 5881 PH for some time, but with *White Bus* the first became L522 MDP and the latter reverted to its original mark L715 FPE, both initially running in the dark blue and gold livery of their former owner.

Former Hodge's MAN (L715 FPE) at Langley Leisure Centre on a swimming transfer for local schools.

The Belgians returned again under the guise of the University of Ghent, though in 2003 they arrived via the Channel Tunnel on a coach to be met at Victoria. Musica Europa once again came to be hosted by St. George's, with the firm providing the transport. The Windsor Leisure Centre ran activity days for children during the Summer break, with coach trips to The Look Out near Bracknell, Wellington Country Park at Stratfield Saye and Beale Park near Pangbourne.

There were covers for *Aldershot Coaches,* as well as for *Yateley Coaches* on their Luckley-Oakfield School run to Wokingham, whilst 'KDT' went on loan to *Reading & Wokingham Coaches* during July 2003.

Following the additional vehicles incoming in 2003 Bedford YLQ-type saloon 'HRO', which had served since it was new in August 1979 departed in October, along with 1975 coach-bodied YRQ 'KDT', which had arrived from *Winkfield Coaches* in 1990. It is pleasing to see that the latter is still preserved, though it passed through various owners before ending up with *Elderson Coaches* of Bromsgrove. Doug had also hoped that 'HRO' might be preserved, letting it go at a low price to enthusiast Lee Simmonds at Sandhurst, but it wasn't to be, and it became a Catering Unit with the BBC at Runcorn instead.

The incoming staff for 2003 only consisted of Frannie Turner as a driver, plus Lionel Davis who was employed as a cleaner.

On the subject of vehicle disposals, it will be recalled that a Children's Entertainer had taken 'GNV' back in 1996. That outfit was run by 'Poz' (family surname Posnet), who was the older brother of someone who Doug had met through the Windsor-Goslar Twin Town exchanges in the 1960's. Some time later he turned up with a bus, not from the firm, but they fixed it for him, leading to the later purchase. In fact he took 'ODL' in late 1996, followed by 'RLW' in 2000 and finally 'EAJ' as The Party Bus in 2004, each of them stripped of seats and with windows covered. He kept them at a farmyard off Riding Court Road over at Datchet, later relocating to Bressingham in Norfolk.

2004-5 – A Return To Dennis

The Royal Borough of Windsor & Maidenhead reviewed its supported local transport and decided to increase the frequency of the *White Bus* Route 01 to provide an hourly frequency with effect from Monday 26th January 2004. At the same time the little used loop around Woodside on certain journeys was dropped, whilst in connection with road re-construction to put a tunnel under Ascot Racecourse, those journeys using New Mile Road between Cheapside and Ascot were also discontinued.

To cover the enhanced service a third DAF SB220 with Optare 'Delta' bodywork was acquired, and it had also originated as an air-side bus at Luton Airport. After that it was converted by *Arriva* at Bradford as a 53-seater, losing its centre doorway in the process, but as that arrangement left no space for luggage, it had the two pairs of inward-facing seats removed in order to give it the same 49-seater capacity as the other pair. As P131 RWR it joined 'GNW' and 'JUA' on the daily rota, which now required 4 drivers to cover.

The improved service was also part of RBWM's new branding of its supported services under the 'Borough Bus' title, which was painted in purple lettering on the body sides and rear, along with the places served.

Optare 'Delta' 'JUA' with Borough Bus lettering, but seen when in use on a Charters School service.

'Delta' P131 RWR with the Borough Bus lettering.

Despite the new branding, there was no inter-working between operators, though of course those knowing their local transport history will recall that *Borough Bus* was an independent covering Windsor to Clewer Green until it sold out to *Thames Valley* in 1955.

On the staffing front Steve Smith came as a driver in March 2004, followed by Richard Bentinck mainly on yard work from November, the latter being the brother of Helen Taylor. On the other hand Joe Sutcliffe retired in April, after which the annual Blackpool outing he had organised was discontinued.

Further transport was also undertaken for The Abbey School on behalf of *Fernhill*, who in turn were asked to stand in on the St. George's trip to Slapton Sands in June 2004. In the early months of that year a number of runs to Southall were covered for Herschell Grammar of Slough, whose own coach and caretaker-driver were not available, whilst some school contracts were run for *E&B Travel (Embling & Barker)* based at Maidenhead.

Pupils who should have passes but failed to produce them were a constant source of difficulty, so from the Autumn Term 2004 a Permit to Travel docket was issued, with a copy to the office to follow up the issue.

Unfortunately the enhancement of Route 01 failed to produce a ridership to justify its retention, so from Tuesday 29th March 2005 the service reverted to basically a two-hourly frequency.

School contracts and related hires still formed the backbone of operations, which in itself did subsidise the continuation of the bus service. However, one of the problems that had to be contended with now were inconsiderate parking at schools, making access and turning difficult, whilst other issues related to the behaviour of students and damage to vehicles. The catalogue of missing window hammers, cut seat-belts, broken knobs and chewing gum were all too familiar to myself when Transport Officer at Wokingham. I remember that one girl actually wrote 'Mica was here' on a seat back on the St. Crispin's Route 39, being

about 13 at the time. In Reception at Shute End Offices they kept packs of coloured crayons and colouring sheets for bored small children, so I took some and 'jumped' the bus, handing her the crayons in front of her peers so she would be occupied on the journey! There were even instances of 'mooning' on the rear of vehicles reported by following cars, whilst the *White Bus* log confirms that the upmarket schools fared only marginally better on such issues. However, one Charters teacher saw the mess her charges had made, so they had to clean it all before going home, and for many years Site Manager Keith Cattran was the Transport Co-ordinator on behalf of that school.

There could also be issues to contend with on adult outings as well, and on a trip to Greenwich for the U3A at Bracknell, one lady failed to return on time. It was then said by another of the party that she might have been seen boarding the *Farnham Coaches* next door, but phone calls found that not to be the case. She had apparently got confused, later turning up at a Police Station, from where she was put on a train from Waterloo to return home safely.

In respect of vehicle changes for the early months of 2005, the former *Fernhill* 1986 Bedford YNV (TIW 2795) with Van Hool 'Alizee' coach body was sold in April, being replaced that month by another from that same source as a 2000 Iveco 'Euro Rider 35', which carried a Spanish-built Beulas 'Stergo E' 49-seater body which included a toilet compartment (W808 AAY), but it only had a relatively short stay until September 2005.

A nearside view of 'Delta' 'RWR' at the Copper Horse on Route 01 to Ascot and via the Royal Lodge.

During 2005 a number of party bookings were taken locally on behalf of the Royal Berkshire Hotel at Ascot and the Holiday Inn at Slough, into Windsor. The Belgian party came again, but that year it was passed to *Fernhill Travel,* as it only required 20 or so seats. The Country Landowner's Association toured the Great Park in June, whilst transfers for the London School of Economics continued to Cumberland Lodge on a regular basis, as did work for Lincoln's Inn, including London to Northampton (Highgate House) throughout the year.

In the Spring and Summer there were more instances of covers for *Courtney Coaches*, which was then having some operational issues, particularly on the Cox Green School contract, as well as some turns on their Crowthorne (Iron Duke) – Bracknell (Ranelagh School) registered service via Pinewood Crossroads.

The WBSEG continued to attend bus rallies and local running days, which resulted in the strange sight of 'JUA' outside Windsor Parish Church on Sunday 17th April 2005 with blinds set for 'Maidenhead 20A', but only as part of the Slough Running Day!

The next vehicle to leave the fleet was another YNV-type, but carrying Duple 'Laser 320' coachwork (D259 FRW), new in 1987 and departing in July 2005. As previously noted there had been some issues regarding seating capacity on certain routes, the 39 to St. Crispin's School in Wokingham being an example, which by predictions for Autumn Term 2005 would need an additional vehicle. However, such figures could never be settled until a short time before term began, as some students were LA-funded and others paid for by parents. If a second vehicle was required, then costs for passes might not reflect that, and about that time Doug confirmed that he intended using a 70-seater then being converted after Military use!

This move also saw a return to Dennis chassis in the form of the 'Javelin', a type introduced to fill the gap left by the cessation of Bedford production, and it had a Cummins C-series straight-6 8.3-litre diesel engine. The first of the former Military examples with bodies by Wadham-Stringer of the 'Vanguard' MkII-type was re-seated for 70 child passengers and re-licensed as L561 ASU in August 2005. It was followed by similar vehicles as L523 MJB in September and L606 ASU in October, all having been new in December 1993 to January 1994.

Former Military 'Javelin' (L561 ASU) is seen after conversion and painting, but prior to being lettered. The rear doors, which allowed stretchers to be carried in its former guise, were not used in its new role as a dedicated school bus, whilst the 70-seater capacity also proved useful for school-related hires.

As well as the ex-Military types, a conventional coach on the same chassis arrived in September 2005 (R674 OEB) with a Berkhof 'Excellence' 57-seater body.

Above - Now fully lettered, another of the 'Vanguard'-bodied 'Javelins' (L606 ASU) is seen at North Street. Note that they lacked destination displays and had painted 'School Bus' signs included in the makeover. *Below –* A nearside view of the type (L523 MJB) with board for Route 83 to Cox Green School.

Courtney's dropped out of the Cox Green contract at the close of Summer term 2005, so *White Bus* was successful as their replacement on Route M03B, and they registered both routes to that school as 83A and 83B from September. The latter started at Haws Hill Farm (Drift Road), then as Fifield Lane – Windsor Road – Holyport – Moneyrow Green (White Hart and Jolly Gardener) – Forest Green Road – Touchen End – Ascot Road – Paley Street - Littlefield Green – White Waltham – Waltham Church – Woodlands Park – Highfield Lane, which covered a route like a large figure-8 lying on its side.

Also from 3rd September *First Beeline's* Route 192 extensions beyond Ascot were re-routed more directly via Brockenhurst Road to reach Sunningdale.

Staff changes at *White Bus* during 2005 saw Josh Canning arrive in May, followed by Phil Chapman and Ken Holloway in September, there also being another Charters route added as the 24R (for 'relief') from Autumn Term 2005.

School-related hires were still the most numerous, with the 'big 3' totalling 168 for Charters, 145 for LVS and 116 for St. George's at Ascot during 2005!

2006-7 – Upping The Tempo

Some issues with vehicles parked at the Bracknell Lorry Park arose from February 2006, when 'MSF' was damaged, the allocation also being 'OEB' and driven by Phil Chapman and Brian Lewis. By May there were 3 kept there and 'DDW' and 'AAO' had fuel stolen and the wheel disks were taken off 'OEB'. However, even the 'home' fleet was not immune, as at 6am one day in March Geoff Lovejoy arrived at the yard to find the doors open on 'Deltas' 'GNW' and 'RWR', the radios having been stolen over-night.

On the staff front Ruth Watson joined as a driver in January 2006, whereas Martin Shaw moved from those duties to be Apprentice Mechanic from April, and Gary Clarke and Sean McAleer joined as drivers in September. In due course Martin would replace Andy Canning as Workshop Manager when he left.

1985 Bedford YNT-type (B542 OJF) with Duple 'Laser' 53-seater coach body was let go in May 2006, whlist other fleet departures saw similar 1983 model (JNM 747Y), but with Plaxton 'Paramount' MkI 53-seater coach body leave in November, followed by the other Duple 'Laser'-bodied example (B30 MSF) new in 1985 out in December.

Berkhof-bodied Dennis 'Javelin' coach (R714 KGK) is shown parked at the Cruchfield Manor Farm out-station. The smart appearance of a mostly white-liveried fleet is a testament to the care given by all the staff at the North Street yard.

Incoming vehicles for 2006 were on Dennis 'Javelin' chassis, one carrying a 53-seater Berkhof 'Radial' coach body (R714 KGK), new in 1998 and arriving in September, followed by another ex-military one with Wadham-Stringer 'Vanguard' MkII 70-seater body (L321 XTC) of 1994, which came in December.

The use of the 70-seaters on school-related hires was quite evident, often where 2 vehicles would have been required previously. However, the situation was over-shadowed by the impending legislation on emissions under Euro 3 and the London Low Emission Zone, as Doug stated in the Trade Press at that time.

Certainly the school-related work was what brought the money in, and hires over the whole age range from pre-schools to senior schools kept the fleet busy. A party from Charters went to The Eden Project in Cornwall for 3 days, and another to Corfe Castle for 4 days, whilst Herschell Grammar School hired 4x53's up to Olympia one day in July.

A shuttle service was operated between Windsor and the Guard's Polo Club at Smith's Lawn on 30th July 2006, whilst a wedding party was taken between the Hilton Hotel in Bracknell, Bray Church and to the reception at 'Englemere' in Kings Ride, Ascot, where the author had his first full-time employment.

First Beeline withdrew all its journeys on Route 192 beyond Ascot after Friday 17th November, though the single school-days run to Charters was retained. This represented the end of the 'Ascot Locals' in various forms since early *Thames Valley* days.

Imperial College (Silwood Park) had a notable series of hires in March 2007, with visits to Northdown Plantation, Ibstone Common and Epping Forest, and the Royal Agricultural College toured the Great Park and Savill Garden in April. Schools visited the Barnes Wetland Centre, and in Southwark the replicas of the Globe Theatre and Sir Francis Drake's 'Golden Hind'. The Eton Dorney Centre hosted a Summer School run by Oxford's Lincoln College, with parties taken out to the Megabowl at Maidenhead, Legoland, Slough Ice Rink and Oxford.

Despite the actions of *First Beeline* back in November 2006, the Company still produced a surprise, when from 26th May 2007 it introduced a 'tourist' Route 300 between Virginia Water (Station) and Windsor! It is perhaps strange that *White Bus* had never attempted such a seasonal link itself, making this the only other service to use the private roads within the Great Park.

The vehicle used was primarily allocated to 'Pegasus' school journeys in the Guildford area, being an Alexander-Dennis 'Super Dart' and DME43922 (LK56 JKE) is seen passing through Ranger's Gate.

The route set out from Virginia Water (Station), then to Savill Garden – Wick Lane – Bishopsgate (where it entered the Great Park), through the Deer Park (with views of Windsor Castle down the Long Walk) – The Village – Prince Consort's Drive – Ranger's Gate – A332 (Kings Road) into Windsor (Parish Church). A tight turn from Bishopsgate into Wick Lane saw the bus on the reverse route travel straight onto Englefield Green, to double-back along the A30 to Savill Garden via Wick Road. Three journeys ran on Mondays to Fridays and four on Saturdays and Sundays, with the seasonal end date of 28th October anticipated.

However, a Foot-and-Mouth outbreak curtailed Route 300 from the 13th September, never to return again in that form. *White Bus* Route 01 was also restricted due to closures, with only Ranger's and Cranbourne Gates left open and fitted with disinfectant straw-mats, so the Deer Park was out of bounds and buses ran to The Village and doubled back, the restrictions being in force until 6th November 2007.

Additional covers ran for *Magpie Coaches* of Slough and *White Knights* of Maidenhead, as well as *Courtney's* contract to Luckley-Oakfield School in Wokingham. The Ascot Wives' Club went to the Oxford Dog-track, enjoying it so much that they went regularly until 2010, always asking for Mick Fazey!

In the meantime incidents at the Lorry Park saw 'DDW' and 'JMJ' broken into in January 2007, no real damage, just things thrown about. In February it was heavy overnight snow that caused problems, and David from *Fernhill* (whose yard was nearby) rang *White Bus* at 6am to advise the roads in the area were too dangerous, and indeed most schools did close that day, but an annoying issue for operators and transport departments was that closures were decided by each Head-teacher and not at departmental level.

In another aspect of Council-related work, each LA kept lists of local firms with specialist vehicles, JCB's etc., to call upon in emergencies, and in July 2007 Neil Beswick of RBWM put *White Bus* on stand-by to evacuate elderly people from Windsor if the River Thames flooded in a threatening fashion!

The Service Van from August 2007 was this Ford 'Transit' (AE51 PKY) from the National Grid (Gas).

The blue 'Enviro' 300 demonstrator is seen posed by a pond near to The Village when used on Route 01.

The matter of suitable types for the service route and school contracts was explored during 2007 with the appearance of a trio of demonstrators. An 'Enviro' 300 'school-bus' 60-seater (SN56 AXM) came from Alexander-Dennis for 29th-30th August, but was not used, followed by a blue-liveried 'Enviro' 300 44-seater bus (SK07 DYA) used on Route 01 on 3rd-4th September. The third example was a red-and-grey bus-bodied Optare 'Tempo' X1200-type (YJ57 EGU), which ran at times between 12th and 15th September.

In respect of staff in 2007, Tony Canning joined in April, followed by Mike Hunt in August and Phil Cook in October.

By the Autumn of 2007 the 1980 Bedford YMT-type (JMJ 633V) was effectively redundant, but it was loaned to *Reading & Wokingham* during October, and on Sunday 2nd December the WBSEG took it out on a 'farewell tour', after which it passed to Andy at *Repton's Coaches,* who intended to preserve it. He also acquired a lot of Bedford spares, but sadly 'JMJ' had to be let go for scrap in due course. Also departing that month was the last Bedford saloon bus (C668 WRT), a 1986 YNT-type with 63-seater Duple 'Dominant' body which had served *White Bus* well

It seems that the Optare bus had been found favourable, and further discussions led to 2 of that type being acquired, though not actually built to order. One was new in November 2007 (YJ57 EHV) and supplied from stock, whilst the other was new in March 2007 and a former demonstrator (YJ07 EGD), the pair seating 47 and 46 respectively. Both arrived in December 2007 and replaced two of the 'Deltas' on Route 01, which in turn took the places of 'JMJ' and 'WRT' on school contracts.

Some of the road cambers soon proved an issue with the 'Tempos' where the Park roads met the public highway, so the Highways Department undertook some urgent work. The third 'Delta' was earmarked as the standby for Route 01 should one of the new buses not be available for service.

2008-12 – Enviros and Volvos

The weather often had its effects on services during the Winter months, but in January 2008 it was fallen trees causing problems. One came down across Dukes Lane in the Great Park, but the driver knew he could go out of Blacknest Gate and double-back to Cheapside, whilst another day one blocked the A332 Windsor Road and 2 *White Bus* vehicles were caught in the traffic. However, a gang of tree surgeons were also in the queue and got to work with chainsaws, and a lorry driver pulled the sawn section clear!

The issue of 'grounding' the lower vehicle types occurred several times when diversions were needed. On one occasions a bus encountered soft muddy grass and had to be towed off by the Park Rangers, whilst in another incident the initial report suggested the fuel tank had been ruptured, but when the van got out with the spill-kit, it was found to be coolant leaking. Generally, any disabled vehicle was fixed on the spot, with Martin or Andy taking a substitute vehicle when required. One coach had a wing mirror broken in London, so another was taken off a coach in the yard and then sent by car to be fitted during the layover.

On the maintenance side, young Oliver Rutter had come as work experience, but basically stayed on to be an apprentice mechanic from March 2008. Brian Cox came as a driver in April, followed by Gerry Bourne as a full-time driver in June, though both had worked there over the years previously at odd times.

Berkhof 'Radial'-bodied Dennis 'Javelin' (V917 TAV).

No less than 4 vehicles came into the fleet in the early months of 2008, with one more Dennis 'Javelin' 57-seater coach carrying a Berkhof 'Radial' body, new in 2000 and seen above, which arrived during January. It was followed in April by a pair of what was to become the new standard choice of coach, Volvo B7R-63's with Plaxton 'Profile' 53-seater bodies (CB53 MTB and DB53 MTB), both new November 2003.

Plaxton 'Primo'-bodied Volvo coach (CB53 MTB).

The 60-seater 'school bus' version of the 'Enviro' 300 had also impressed when it visited earlier, and after *Courtneys* merged the Holt School 62a/62b routes onto a double-decker, the opportunity came to buy one of its 2005 examples (SN55 DVB), which came in yellow 'Cool Bus' livery in April 2008. It was duly joined by its batch-mate (SN55 DVA) in May of the following year when it too was released.

'Cool Bus' (SN55 DVA) waits at Charters School.

Displaced by the incoming vehicles in March were a Bedford YNT-type (D633 DDW) and a MAN 18.370-type (L715 FPE), followed by the last Bedford in the fleet, an YNV-type (E849 AAO) in June, and then the other MAN (L522 MDP) in September. The diary for July 2008 also noted that the 70-seater 'Javelins' must not be used in the London Low Emission Zone.

In the meantime the National Concessionary Travel Scheme came into effect from 1st April 2008, with those attaining the State retirement age (for women) could apply for a Pass, along with those eligible for the previous Disabled Travel Pass, funded by the Department of Transport and Local Authorities. There is no doubting the value of the freedoms the pass has given many people who might have otherwise been isolated, whilst the degree to which marginal bus services have survived cannot be under-estimated, even though the debate on reimbursement continues, the cost of alternative social transport is much higher!

At Easter 2008 *Courtney Coaches* withdrew from its Luckley-Oakfield school run from Cookham, so it was offered to *White Bus* through to the end of the Summer Term. Starting at 7.40am at the Station, it ran as Cookham Dean (Church) – Maidenhead (All Saints) – Highway Avenue – Twyford (Waitrose and Station) – Hurst (Pond), with a return journey from the school at 16.10. However, when the full contract was offered for the Autumn Term 2008, it was not taken on, as the late coach at 17.40 did not fit in with other commitments, and it instead went to *Yateley Coaches,* and then later to *Reading & Wokingham.*

As it was, from September 2008 there was an additional school run, the second as Route 08, worked off the back of the St. Crispin's 39 from Wokingham to LVS in Ascot. In the end only two points had any assignments, at Forest Road/Foxley Lane and at opposite the Royal Standard at Binfield Village, and it ceased after only about a year due to lack of numbers.

In the Slough area there were still more additions to the swimming transport, with Dair House, Lent Rise, Stoke Poges and Long Close all added during 2008.

RBWM prompted another re-launch of Route 01 from Monday 26th January 2009, creating a new image with vinyls depicting the main attractions served, with the 'Tempos' kitted out for the press event at Ascot Racecourse and at The Village. However, as the design used 'contra-vision' panels over half of the windows, it rather defeated the object of admiring the view, and indeed adverse comments were received!

One of the 'Tempos' (YJ57 EHV) in the Borough Bus livery, but for photos of these in the new vinyl scheme see page 126 of the colour section.

There were 4 wedding jobs during 2009, whilst one lady liked the yellow 'Cool Bus' livery so much she hired one for her Diamond Wedding event in the Park in April. Also there was the WINGS encampment once again, and a coach collected from Heathrow the Girl Scouts of Western Washington in August.

Several shuttle services were operated, the first for the Jack Wills Polo Day on 6th June between Windsor and the Guard's Polo Club at Smith's Lawn, which was repeated for 2010/11. The other was for the Victorian Fayre held in Sunninghill High Street on Sunday 15th November, when a bus ran between Car Park 3 off Ascot High Street and Sunninghill, provided at a good rate to what is largely a charity event, and that has been repeated each year since, with a route board too!

In 2009 Roy Annetts was the only new driver, starting in January, whilst the only departure from the fleet was the first of the 'Javelins' to be sold (L321 XTC), which left in December.

One of the 'Tempos' (YJ57 EHV) in snowy Windsor.

The year ended with heavy snow on 21st December, so the Route 01 ceased after the 16.15 from Windsor, and the road conditions were so poor that the *White Bus* party had to be cancelled. 2010 started with some 6 inches of snow on the ground, and most schools did not re-open for the first week of January. Even after that many school trips were cancelled.

Various vehicles were inspected with a view to making them LEZ-compliant during 2010, though it was concluded that the Volvo B7R's were the only suitable candidates, after which they were fitted with 'Eminox' equipment for use on any London hires.

For the Summer break 2010 LVS hosted foreign students under the 'Skola' and 'BCS' schemes, in similar fashion to previous work by *White Bus* for Nord Anglia, with parties from Spain, Italy and Russia coming over in groups during July and August.

A third Route 08 started from September 2010, after the LVS closed the Elvian School with which it had replaced Presentation College at Calcot. Therefore a run left Calcot (Sava Centre), then via Shinfield Park – Binfield (Temple Way) – Harvest Ride (Bracknell to Warfield) – County Lane, which ran until July 2012. A further Charters School variant added from Autumn 2010 was the 24D.

Parking at Longshot Lane ceased with the end of the Autumn Term 2010, the 3-vehicle allocation going to join others at Cruchfields Manor Farm instead. The year once again ended with snow, so Route 01 was taken off after the 11am ex-Windsor took 2.5 hours!

We will now step aside from the chronological events to see how the day-to-day management of operations evolved during the later years. After Jim Waterfall came off driving duties due to ill health in 1988 he took on various tasks in the office, with arranging hire work, book-keeping and started 'The Board', where commitments were chalked on a blackboard. Helen Taylor joined in October 1998 as Book Keeper, then over the years she took on more tasks, such as the issue of scholar's season tickets, and with other changes in 2006 the title Office Manager came about.

Although Doug's wife Jan had not played an active role in the business, bringing up their 3 children and of course giving him support, her diagnosis of cancer in 2005 did give a focus to the need for more time for the couple, and the delegation of the workload. To that end Geoff Lovejoy, who had joined as a driver in 1993 and Phil Bathard, also a driver from 1998 were made up to Operations Supervisors from 2006.

In the meantime Helen's brother, Richard Bentinck, had returned from living in Spain and did work on the garage in 2004, before starting part-time later that year helping out with the maintenance and yard, then later as full-time and gaining his PSV License in 2005. In due course he took on the Health & Safety role and from 2006 was designated Yard Supervisor. After Geoff left again in May 2013 Richard became the other Operations Supervisor in his place.

Sadly Jan died in September 2009 aged just 57. Doug's brother Gerry Jeatt has also helped out at times with 'number-crunching', whilst each of the children has seen spells in the office, with daughter Claire now employed two days a week helping Helen.

From April 2011 there was another variant to Charters School as the 24E. Additions to the swimming runs for schools saw the Slough-based Iqra, Khalsa and Oldfield Schools covered from January 2011. The Bullbrook-based Holly Spring School went to LVS weekly, also by *White Bus* from that July for swimming lessons.

June 2011 saw a shuttle service operated from the Bagshot Road to the Motor Trade charity BEN's annual fete at Lynwood, on The Rise at Sunningdale, repeated each year since for the Saturday event.

White Bus staff also attended the funeral on 5th December of former driver Bob Napper, and a wake afterwards in Chalvey at The Garibaldi, a pub he often organised excursions on behalf of over the years.

There was no staff additions for 2011, but 4 vehicles arrived and 2 departed. In October another 'Enviro' 300 60-seater 'school bus' was acquired (MX09 GZL) and it came from the in-house fleet of West Sussex CC. New in 2009 it had a yellow and black livery, which it carried with *White Bus*. Also acquired that month was another Volvo B7R-63 with Plaxton 'Profile' coach body with 70 seats (LD04 MCT), new in 2004. The following month another pair of Volvos came (YN10 ABK/YN10 ABO), new in March 2010 and seating 57, that type being suitable for use in the London LEZ. *A photo of 'GZL' appears on page 127, with 'MCT' and 'ABO' both on page 148.* The outgoing vehicles in November 2011 were both Dennis 'Javelins' with Berkhof 'Radial' bodies, R714 KGK with 53 seats and V917 TAV as a 57-seater.

No fleet changes took place during 2012, but January saw drivers Warren Barry and Gavin Fenton starting. Accommodation was still an issue at North Street, so 2 vehicles were out-stationed a mile to the west at Cheval Stud, off Pigeonhouse Lane from March 2012.

From Thursday 22nd March Route 01 was amended in Windsor to operate direct from a stop in Charles Street to Clarence Road and Victoria Road on the outgoing journeys, no longer looping through the lower end of Peascod Street as it was closed off to traffic.

First Beeline introduced yet another tourist service with Route 100 from Saturday 2nd June 2012, which covered a circular route from Windsor (Riverside Station) – Great Park (Kings Road) – Cheapside – Virginia Water (Wheatsheaf) – Egham – Runnymede – Old Windsor. It was advertised to allow stops off en route, in particular for the Magna Carta, RAF and Kennedy memorials at Runnymede, with a descriptive leaflet and map. Some Volvo 'Olympians' were re-painted in a mostly red, white and blue scheme. Indeed, the buses proved rather unreliable, with numerous failures in service! It did not enter the Great Park, and it ceased on 26th October 2012.

First Beeline Volvo 218 (S218 LLO) at Windsor.

A Transport of Delight
By Terry Dwan

**Delivered on the occasion of
a family gathering on
1st September 2012**

'Some people like to use car
A train or motor bike
As easy ways to get around
And ones they rather like
Such means of locomotion though
Seem very dull to us
Third generation owners of
The Winkfield Omnibus

Hold very tight please
Hold very tight please

When you're lost in the Great Park
And don't know where you are
You could do worse than flag us down
The journey won't be far
And very soon you will be at
The North Street terminus
Of that monarch of the road
The Winkfield Omnibus

Along the Berkshire highways
We drive our merry load
At 30 miles per hour
In the middle of the road
We love to drive in convoys
We're most gregarious
That streak of bright white lightening
The Winkfield Omnibus

When natives try to pass us
Before they overtakes
We stick our big right hands out
As we jam on all our brakes
That sure does get them livid
And really gets their goad
As PSV's we're immune from
The British Highway Code

Doug don't ask much for wages
He merely wants fair shares
So he's cutting down the stages
And stuffing up the fares
If tickets cost £10 apiece
Why should we make a fuss
It's worth it just to ride inside
The Winkfield Omnibus'

Terry is a cousin to Doug Jeatt,
and he acknowledges the original
item of the same title by the
comic songwriters Flanders &
Swann as his inspiration.

Seen at the Cruchfield Manor Farm out-station at Hawthorn Hill are Optare 'Delta' bodied DAF (S158 JUA) and Plaxton 'Profile' bodied Volvo B7R-63 (LD04 MCT) with 70 seats for school work in March 2013, quite typical of the allocation mix at that time.

Above – Chris Bevan of Bevan Signs applies the vinyl lettering on one of the coaches. **Below** – The completed lettering as seen on Volvo B7R-63 (YN10 ABO) with Plaxton 'Profile' 57-seater body parked alongside the North Street yard.

There was also a new swimming contract from early 2012 for schools in the Maidenhead area, with Alwyn, Lowbrook and St. Peter's pupils taken to the Magnet Leisure Centre in Maidenhead, and from April Crazies Hill was added, followed by Boyn Hill, Cookham Rise, Holy Trinity, Larchfield and Knowl Hill from September. From the Autumn Term 2012 LVS Route 09 included Crowthorne and Sandhurst each day.

2013-15 The Windsor Locals

From 7th January 2013 another existing LVS service was taken over as Route 06, which runs from Chalfont Common via Chalfont St. Peter – Gerrards Cross – Stoke Poges – Slough – Windsor, which had been previously covered by *First Student, Fernhill* and *London Country*. The latter could not alter timings to suit term ends, so the school gave it to *White Bus!* On the subject of Licensed Victuallers School, there was also a single-day only operation at the start of Autumn Term 2013 from Chertsey via Woking and Chobham, but the in-house minibuses were available again after staffing issues were resolved. From the academic year 2013-14 Route 08 from Calcot ceased.

Further changes to Route 01 from 1st July 2013 were to accommodate the Crown Estate's desire to limit vehicular traffic through the Deer Park, so journeys onto Royal Lodge now took a detour via Dukes Lane and Peewit Gate to get reach The Village. It also saw the end of journeys via Forest Gate and Fernhill, with all runs except the positioning journey in the morning and the fast run to Windsor at 17.15 now entering the Park through Ascot Gate, with other points now only called at by request or pre-booking.

Staff changes in 2013 saw Simon Platt join in May, Joe Curly in August and Paul Fenton in September.

Regarding the fleet, 2 more 'Javelins' departed, with Berkhof-bodied coach (R714 OEB) in January, and Wadham-Stringer-bodied (L606 ASU) in August. A further pair of Volvo B7R-type coaches came in July (YN59 GPE and YN59 GPF), both new in October 2010 with 70-seater Plaxton 'Profile' bodies.

Volvo (YN59 GPF) is seen at North Street Garage.

Several rounds of tendering of Local Bus Services and School Contracts for RBWM saw a number of changes affecting *White Bus*. Co-operation with those in the Transport Unit ensured several gaps were filled in the Holyport area, where numbers had reduced but *White Bus* agreed to continue with Route 83 to Cox Green and Route 88 to Windsor schools. From the opposite direction, Route 13 was extended on from Old Windsor to Windsor, but now as part of Route 01.

The Local Bus tenders saw *White Bus Services* gain several local routes in Windsor, each of which already existed. One was a daily West Windsor Local Service as Route W1, which had previously been provided in a limited fashion by *People to Places,* itself in urgent response to reductions in the Dedworth and Clewer areas by *Courtney Bus* and *First Beeline* earlier. It ran from Castle Hill via Charles Street – Addington Close (Clewer) – Sebastopol PH – Dedworth (Tinkers Lane) – Blackhorse PH - Vale Road – Smiths Lane – Gallys Road – Ruddles Way – Blackhorse PH – Bell View (Clewer) – Addington Cl. – opposite Parish Church. It had 5 off-peak journeys and operated on Mondays to Fridays, with flat fares of adult single at £2.20 and adult return at £3.60, along with child fares.

For the first few weeks of operation other vehicles were used on the Windsor locals, as the new buses ordered were delayed. Chris Spencer caught one of the 'Deltas' (R89 GNW) on Route W1 at Windsor.

The other daily service was a Park-and-Ride format, mostly operating between the Home Park and King Edward VII Car Parks to the east of the town and to Castle Hill basically with a 20-minute headway. However, as with the previous *Courtney Buses* Route M4, this Mondays to Fridays operation featured 3 journeys extended from the Home Park onto Datchet. On reaching Datchet the bus then made a circle of the village and its main housing estates via St. Mary's School – Ditton Road – Major's Farm, to then return to The Green, where it terminated prior to returning the direct route over the railway line and the River Thames to call at King Edward VII Car Park and into Windsor Town Centre. Known as Route P1, the bus has to travel over an excessive number of speed humps in the Home Park Car Park, but passengers use the service for free other than paying for car parking.

The third operation was another Park-and-Ride, but only running about 40 days each year, with the Easter weekend, weekends in July and August, and then for the pre-Christmas period weekends from November to December once the Christmas Lights had started. As Route P2 it replaced directly the service contracted to *Courtney Buses* as M5, and when running it started at Charles Street and took a circular route via Clarence Road – Windsor & Eton Relief Road – Maidenhead Road outwards to the car park of Centrica at Clewer, returning along Maidenhead Road – Arthur Road – Ward Royal to Charles Street, and operating every 15 minutes between 10.00 and 18.15. Again, use of the bus is free, with car parking charges made, whilst the English National Concessionary Pass is also valid.

A pair of 'slim-line Optare 'Solo' M790SE-types with 27-seater bodies was ordered to cover Routes W1, P1 and P2, whilst a new pair of 11.7-metre long Optare 'Versa' V1170-type 44-seaters was ordered to take the place of the 'Tempos' on Route 01. However, when the Windsor locals started on Monday 3rd February 2014 none of the new buses were available, so a pair of Hungarian-built Enterprise Bus chassis with Plaxton 'Primo' bodies was hired from Dawson Rentals, being KX57 FKW with 28 seats and KX57 OVC with 22 seats.

In another Chris Spencer photo we see one of the hired buses 'OVC' on Castle Hill on Route P1.

The P1 got off to a bad start when Datchet and the two car parks suffered badly from flooding by the River Thames and was suspended for several weeks, whilst echoing that bad luck the bus on the first day of P2 was reversed into by a car! More drivers were taken on for February, Krippasur Rai and Ishwar Limbu, coming from *Courtney* on the Windsor locals, joined by Richard Heron and fellow Old Windsorian pupil Robert Turner, who already had a PCV License.

There might have been even more new routes, as over at Bracknell a falling-out between the Council and *Thames Travel* meant that an interim tender was made available in January 2014, but whilst Doug did consider the possibility of bidding for those routes out to Winkfield, he decided there was enough going on.

One of the 'Solos' is parked out at Trevelyan School off St. Leonard's Road, and seen here is YJ14 BBN. A photo of one of the 'Versas' appears on page 128, along with other 2014 views in colour.

The new buses started to arrive, with the 'Solos' (YJ14 BBN and YJ14 BBO) entering service from 30th March, followed by the 'Versas' (YJ14 BBU and YJ14 BBV) from 16th April, and as a result of duty changes Optare 'Delta' P131 RWR was sold in June. That left the fleet as 20 vehicles, which are those marked as 'current' on the Fleet List on pages 158/9.

From July 2014 a single-journey daily shuttle has been operated on behalf of *Courtney Buses* between Langley Station and Cisco Systems at Bedfont Lakes near Feltham, whilst the latest additional LVS service was that taken over from in-house minibuses as Route 04 in September 2014, starting in Maidenhead (Boyn Hill) and via Bray Road – Holyport – Moss End – Newell Green – Warfield (Tesco).

More new drivers were Darren Ford In September 2014 and Paul Styman in October, whilst Troy Upton started in February 2015. In July 2014 David Bentinck (son of Richard) started as Apprentice Mechanic, the third to be trained over the years, as well as a large number of drivers for such a small operator.

Changes had to be made to out-stationing from June 2014 after the pair at Cheval Stud needed to leave, so they are now at Buckhurst Moor, near Amen Corner in Binfield, which cover the St. Crspin's 39 journeys.

On 19th December 2014 the firm's longest-serving driver, Mick Fazey, retired after 31 years as full-time, and heaven knows how many miles, appropriately dubbed as 'Mr. SatNav' by one of his colleagues!

Winkfield Coaches 1955-1990

The background to operations from the North Street Garage has already been fully explored under the main text, as have the relevant changes to legislation and other local factors. This chapter therefore sets out to further examine the background to Dick Mauler and the developments of the *Winkfield Coaches* operations and fleet for the 35 years of its separate identity.

We have of course heard quite a bit about Vivien's earlier days, helping out with the business in various ways. She got her first license to drive a car in 1928 at just 17, in an age when very few women drove, and in fact of those most of them were the 'bright young things' with rich parents. Vivie has also been noted as the 'first woman bus driver in the South' and, whilst not strictly the case (as there were others before her) it does seem possible she may have been the first licensed for PSV's under the 1930 Road Traffic Act on 2nd October 1931, Driver's Badge J4002, along with Conductor's Badge J3045. After the disbanding of the large Southern Traffic Area, she was issued with the first of her South Eastern badges in 1934 as K1623 for driving and K7603 when conducting.

Dick Mauler all dressed up and ready for the off.

'Dick' was born as Eric Mauler in April 1915 at Avening in Gloucestershire, where by 1911 his Father was a domestic coachman. The family then moved to Sunninghill, though later back to Stroud, but Dick stayed and by the late 1930's he was at the Broadlands Estate off Bagshot Road in Sunninghill, looking after the show-horses. During that time he first met his bride-to-be Vivie Jeatt, and in the war years he served in the RAF as a Corporal, riding a motorcycle as

escort to the large Queen Mary Transporters taking aircraft all over the country.

The initial pair of vehicles taken out of *White Bus* in September/October 1955 was a 1939 Bedford WTB-type (BUN 677) with Duple 'Hendonian' 26-seater front-entrance coach body, and a Harrington 30-seater front-entrance bodied 1948 Commer 'Commando' 17A-type coach (EUX 524).

Above – Bedford WTB-type (BUN 677) at Goodwood, and *below* – Commer 'Commando' (EUX 524) with a previous owner.

Another similar 1946 Commer, but with a Plaxton K3-type 30-seater front-entrance coach body (GOM 658) was added from a West Bromwich operator, one source quoting October 1955, but believed to be a few months later than that and via others.

A Bedford OB with 29-seater Duple 'Vista' body (JXH 719) came later in 1956, ousting the 1939 WTB, which also came via others, but had originally been in the London-based *Fallowfield & Britten* subsidiary of *Grey-Green*.

The Commer 'EUX' was replaced in 1957 by a rare Maudslay 'Marathon' MkIII (SMU 212), that chassis maker about to disappear into AEC, and it carried a 35-seater fully-fronted Plaxton 'Crusader' body with front entrance, another vehicle of London origins in 1951 latterly with *Valliant* of Ealing.

Vivien Mauler was also active behind the wheel, with her main task being the schooldays Charters contract from the local area. For 20 years she used a 38-seater Duple 'Vega' bodied and petrol-engined Bedford SBG (BJP 76), which was acquired in 1957 and sold in 1979. When new it was with *Sexton & Blagg* of Widnes, but came via another operator.

Vivien Mauler with 'her coach' (BJP 76), outside the family home in Crouch Lane with Patricia ('Tish').

Another Duple 'Vista' 29-seater-bodied Bedford OB (BDC 387) followed in 1961, which replaced 'JXH'. It was new in 1950 to *Ayrton* of Grangetown, but also came via others, and was noted for spending much of its time being on hire for weeks at a time for a local film studio, being rarely seen back at the base. On the other hand 'JXH' only strayed a few yards to *White Bus* for further service with Cecil Jeatt!

As to the livery used, by the early 1960's John Gillham recorded a cream and maroon preference.

AEC 'Reliance' (MOW 501) when with Hutfield's.

Next came an AEC 'Reliance' MU3RA, which at that time featured a 7.865-litre diesel engine, and this one (MOW 501) carried a Burlingham 'Seagull' 41-seater front-entrance coach body. Acquired in 1962, it had been new in 1954 to *Hutfield's* of Gosport. That was followed in 1964 by a similar 1959 example, which carried an early Plaxton 'Panorama' body with seating for 41 (208 VHX). It had been new to *Contravel* of London EC2 on high-class tourist work, and it was the replacement for the Maudslay 'SMU'.

The Maudslay 'Marathon' coach (SMU 212).

Another of the recollections by Michael Clancy notes that Dick Mauler looked quite small when at the wheel of one of the 'Reliances', whilst he had said that he found them heavy to drive after the Bedfords. That comment aside, the type was present in the fleet from 1962 to 1978.

The basic work pattern did not vary much over the years, with the bread-and-butter tasks being the daily worker's contract from The Squirrel to the Rathdown Industries light-engineering works at London Road in Ascot, along with the Charters and Wick Hill (Bracknell) School runs. There was also a run for J.T. Engineering of New Road in North Ascot used by them to bring in skilled workers until buying a former *Maidstone & District* Beadle-Leyland rebuild (OKP 984) for that purpose in the early '60's. From Orchard Lea, at the junction of Winkfield Lane and the Drift Road, a firm making garments from 'Lurex' metallic yarn had workers brought in, and later SGS Inspection Services had a contract from Camberley to that spot.

Apart from that in the earlier days there were still pub-related hires, along with other functions and outings, though no licenses were held for advertised excursion work. From an early date there was a twice-weekly shopping run into Windsor or Bracknell for those not on a bus route, which gave a couple of hours in town and fitted in nicely with other commitments. As work had developed the original man-and-wife team of drivers was supplemented by hired men, with Jack Edwards, Les Spong and Joe Sutcliffe all joining.

Les Spong was very much a local man, born in 1935 at a cottage in the Great Park, and still living there to this day. His Father was a worker in the Park, and used to help out fuelling and cleaning vehicles at North Street. On leaving school Les became a fitter-engineer on the agricultural side, and went for his National Service in 1956/7. On returning he found several of the lorry drivers had also obtained PSV Licenses for part-time driving with *Winsdorian Coaches*. Encouraging him to do the same, one of them pointed out that Dick was looking for another driver, so he came as part-time in 1957. He worked for Doug Jeatt from 1990, but after a fraught job up to London in 2006, decided at 71 to call it a day!

Joe Sutcliffe had worked as a fitter at Bracknell for *Thames Valley* from 1960, then with *Moore's* and *Lynwood Coaches* of Old Windsor, before coming to the North Street with TV driver Jim Collins to help maintain the vehicles, when Dick Mauler duly offered him a job driving, whilst Jim Collins took up the same offer with *White Bus*. Joe was also active in arranging various outings, but his annual speciality was an extended trip to view the Blackpool Illuminations each November, something he continued with after he also transferred in 1990. When *Winkfield Coaches* hired 'RLW' to *White Bus* it was as 'Joe's coach'.

There were also other part-time drivers at times, Tim Webb, Rod Haylor and Ted Waterman, whilst TV Inspector Harry Stone, who lived in North Street with his French wife, shared the driving with Dick on the annual Charters extended trip to Scotland. Harry later worked in the yard most days up to his death at 85 in 1987, his parents having run The White Hart at Winkfield by 1911 and being there for many years.

Joe also recalls how Dick could be as filthy as anything in the garage, but before going out on a job would get all scrubbed up, donning a white shirt, tie and suit, earning him the nickname 'the smart man'. His daughter Christine also recounted how earlier on he used the stiff detachable collars, then rather out of fashion. After he could no longer obtain those he settled on 'Van Heusen' shirts due to their quality. The Gloucestershire contacts were kept up by Dick's sister Ursula and her husband Don, licensees of the Dukes Head in Sunninghill by organising a pub outing to Stroud and calling in to see their relations.

For 1967 another new type was tried as a replacement for 'BDC', it being a Ford 570E (8122 PU) of 1960 with a Plaxton 'Yeoman' 41-seater front-entrance coach body, which came after 4 other owners. That was also the year that *White Bus* acquired a similar coach (YFH 53), but it was short-lived at only 2 years. Of course the closeness of the two firms and shared maintenance facilities resulted in a number of similar types featuring in both fleets. A further AEC 'Reliance' of the 2MU3RA type (YVA 871) followed in 1969 to replace the Ford, carrying a Duple 'Britannia' 41-seater front-entrance coach body and new in 1961 to *Irvine* of Salsburgh but latterly with *Green Coaches* of Walton-on-Thames. Some of the vehicles purchased by *Winkfield Coaches* (and *White Bus*) in the 1950's to '60's came from Percy Sleeman the dealer based in Ealing, London W5.

AEC 'Reliance' (YVA 871) as new to Irvine's.

However, one Bedford type never to feature in the *White Bus* fleet was the 3-axle VAL, of which the first example arrived at *Winkfield Coaches* in 1971. With its twin-steer arrangement and 16-inch wheels the model gave a very low floor height at a time when the under-floor engined types did quite the opposite. For that chassis Plaxton had created a special 'VAL' body with seats for 52, though with the front engine position still retained, that caused the layout to include several single seats on the nearside towards the front. This example (BOO 165B) had been new in 1964 but came via others in order to replace the AEC 'MOW'.

Bedford VAL-type (BOO 165B) after sale to Porter of Dummer was photographed by Mervyn Annetts, and which wore a two-tone blue livery.

This earlier example was the VAL14, with a Leyland 0.400 6.17-litre diesel engine, and whilst the type was popular as one of the first high-capacity coaches, the earlier examples were prone to the linkage in the twin-steer coming adrift, leaving sets of wheels pointing in different directions! Another issue as noted by Les Spong was the tendency for more frequent punctures, as the lead wheels shot any road-based items straight into the trailing set instead of to a wheel arch. Despite any drawbacks the VAL was represented in the fleet from 1971 to 1984, and of course the type will also bring to mind the film 'The Italian Job'.

Another updated version of the AEC 'Reliance' with more powerful AH505 engine (HAR 213C) came in 1972, new in 1965 to *Sapphire Coaches* of London WC2, then taken over by the succession of *Valliant Coaches, Valliant-Cronshaw* and lastly *Venture,* in whose distinctive yellow and black livery it arrived.

The first new purchase by the firm arrived in September 1972 (GUR 416L), as a Bedford YRQ-type with Duple 'Viscount' 45-seater front-entrance coach body. It should be noted that the following month the same combination was delivered to *White Bus* as LJB 403L, but each would last only just 7 years.

Bedford YRQ-type (GUR 416L) emerges into the High Street from Windsor Central Station. To the left is the hotel variously known over the years as Ye Harte & Garter and the White Hart, which also has its place in local transport history. Before the British buses came to the area that hostelry also ran a motorbus service from the town to Old Windsor, and some charabancs.

After that there was a run of 9 more Bedfords of 8 different chassis types and featuring 8 body styles, the fleet comprised solely of that make from 1978.

1973 saw a Bedford SB1-type with 41-seater Plaxton 'Consort' MkIV front-entrance bodywork (4402 NX), new in 1960 to *Lloyd* of Nuneaton but via others. It was joined that year by a 1968 Bedford VAL70-type, which featured the Bedford 7.6litre diesel engine, and this example (HTF 449F) carried a 'Viceroy 36' body with 52 seats built by Duple (Northern) at the former Burlingham Works at Blackpool, being delivered new to *Robinson* of Great Harwood. These two vehicles replaced 'BJP' and 'BOO' of similar seat capacities.

In respect of developments with a standard livery, at least a couple of vehicles (BJP 76 and 4402 NX) were photographed having the coloured areas replaced by a turquoise shade of blue-green, though the cream paint was left unaltered, those dating from 1971. However, apart from that, and the addition of maroon to the all-over white of 'MOW', *Winkfield Coaches* no longer

pursued a practice of repainting as time went by, and vehicles therefore ran in a variety of liveries.

'HTF' only stayed for a year before being replaced by the same type (PLG 702G), but bearing a 52-seater Plaxton 'Panorama Elite' front-entrance coach body, new in 1970 to *Bostock's Coaches* of Congleton, in whose fawn and red livery it continued to be used.

From rather closer to home, the next to arrive in 1975 was former *Windsorian Coaches* 1963 Bedford SB8-type with Leyland 0.350 5.76-litre diesel engine and Duple 'Bella Vega' 41-seater body (81 DBL), which ran it the grey and red livery. 1976 saw another Y-series Bedford, a YRT (TCG 333M), new just 2 years before with a Duple 'Dominant' 53-seater body and acquired after service with *Coliseum* of West End.

Of course, it must be appreciated that a small operator with such limited facilities for bodywork overhauls and painting tended to buy a vehicle with a few year's Certificate of Fitness and then sell it once work became an issue. That probably also explains why a number of the vehicles ended up with youth groups in non-PSV service, with BJP 76 going to the 2nd Harrow Sea Scouts, BDC 387 to the Redlands Youth Club at Coulsdon, 4402 NX to the 4th Woking Scouts and 81 DBL staying locally with the 2nd Bracknell Scouts. In fact the latter stayed even closer to the fold, as it was Jack Edwards who now drove it as Scout Leader!

Outgoing in 1975 was AEC 'Reliance' (208 VHX).

The final quartet of purchases saw a 1975 Bedford YRQ-type with Plaxton 'Panorama Elite' MkIII 45-seater body (KDT 281P) arrive in 1979 via several owners, but new to *Globe* in Barnsley, then in 1980 a VAL14 with Plaxton VAL 52-seater body (52 BDL, the registration to celebrate being the first coach of that capacity on the Isle of Wight so it is said), new in 1963 to *Moss Tours* of Sandown. This coach ran in the two-tone blue of its original owner, whilst it should be appreciated that although 17 years old, there

Bedford YMT-type 'RLW' was caught by Phil Moth in the side streets of his native Camberley on the occasion of the large Royal British Legion parade held each year at the Royal Military Academy.

was relatively low mileage as many coaches were stored for the Winter months on the Island. It replaced 'DBL', being joined in 1981 by a YMT-type with 53-seater Duple 'Dominant' body (RLW 778R), which replaced 'TCG', and was new in 1977 to *Fox* of Hayes, Middlesex. The final arrival was in 1983, with another Plaxton 'Panorama Elite' MkIII-bodied YRT, but with seating for 53 (GHD 668N), and which despite its Yorkshire mark started out with *Silver Fox* of Edinburgh.

In the meantime Vivien Mauler had passed away in January 1980 at the age of 69, having been involved with the businesses at North Street Garage since her teenage days. Dick Mauler re-married in late 1981 to Janet Taylor, who had come from Reading originally, but via Market Lavington in Wiltshire. She had joined the staff at Charters School in September 1959 and risen to Head of Biology, so Janet and Dick knew each other through the coaching activities at school.

In 1984 Christine was given some PSV training by Joe Sutcliffe and gained her license to help her Father out, though not doing much driving as she had a daytime job at Barclays Bank in Windsor. However, that also brought about an interesting regular task for the firm, with the bank ringing at short notice with a very precise time for a coach to stop outside the main

branch at the top of Peascod Street. As it pulled up staff would come out with bags of money to be taken to the Datchet branch, about 1.5 miles to the east.

During early 1984 the last example of the VAL's was sold (52 BDL), leaving the trio of Bedford Y-series as the fleet through to the end of operations.

During 1990 Dick decided to finally retire, offering the business to his nephew Doug Jeatt, who was able to finally re-unite the businesses at North Street. Dick moved to Janet's house a few miles to the west in Warfield Street and he passed away in 1995 shortly before his 80th birthday.

Bob Mack found Bedford OB (JXH 719) on the White Bus service at Windsor Central Station after its sale.

Some White Bus Drivers – *Clockwise from top left –*
Alan Moore, Ron Coxhead, Ken Acton, Joe Sutcliffe,
Greg Edwards and Chris Barber.

White Bus Staff 1930 - 2015

This list has been compiled from all available sources, as (hopefully) a complete record of all those employed by White Bus Services as drivers, maintenance and yard duties, as well as office staff over the years.

Representing some 150 years of service between them are (left to right) Mick Fazey, Ron Richmond, Geoff Lovejoy and Doug Jeatt in July 2002 at North Street.

Don ?	Vivien Jeatt	Cecil Jeatt
Sid ?	Maurice Nugent	Jim Waterfall
Will Jeatt	George Jacobs	Rod Haylor
Ray Bartrop	Wally Freeman	Ray Cleeton
Chris Barber	Joe Sutcliffe	Alan Moore
Les Spong	Geoff Lovejoy	Mick Fazey
Ted Waterman	Allen Titchenor	Ken Cooper
Ian Dundas	George Lynham	Ron Coxhead
Brian Lewis	Tony Davidson	Josh Canning
Graham Messenger	Les Birchmore	Ken Holloway
Ray Shears	Michael Poynter	Phil Chapman
Graham Ottewill	Phil Cook	Bob Napper
Brian Lovejoy	Richard Ludlam	Terry Dwan
Frank Hamilton	Jim Collins	Tony Smith
Barry Devaney	Franie Roche	Reg Brooks
Richard Bentinck	Andy Canning	Barry Brophy
Peter Horton	Greg Edwards	Eric Smith
Gerry Bourne	Terry Gibbs	Helen Taylor
Ron Richmond	Mike Smith	Mike George
Eric Chambers	Joe Curley	Bob Pratley
Lionel Davis	Mike Hoare	David Stokes
Warren Barry	Phil Bathard	Steve Carlow
Bill Darling	Carol Venables	Brian Cox
Tony Canning	Iain Hurley	Martin Shaw
Ken Acton	Francis Taylor	Gary Robins
Dave Empson	Chris Empson	Mike Swift
Ruth Watson	Carol Masters	Len Wright
(later Soden)	Gary Clarke	Mike Hunt

Oliver Rutter	Les Copley	Sean McAleer
Mike Groves	Roger Moore	Peter Ives
Roy Annetts	Ian Harris	Darren Ford
David Bentinck	Paul Felton	Paul Styman
Richard Heron	Steve Smith	Robert Turner
Tony Stockdale	Francis Turner	Krippasur Rai
Ishwar Limbu	Troy Upton	John Hughes
Ray Sutherland	Gavin Fenton	Doug Jeatt
Ashley Titchenor	Simon Platt	Gerry Jeatt
Katherine Taylor	Sally Edwards	Claire Jeatt
Nicola Lovejoy	Dick Mauler	Matt Jeatt

Above - Another group in the yard shows (left to right) Geoff Lovejoy, Chris Barber and Jim Waterfall with WPM. Below - Taken on the occasion of Alan Moore's retirement in May 2000, here we see (left to right) Mike Smith (aka 'Stitcher'), Brian Lovejoy, Ken Gagan (not staff, but a member of the WBSEG), Les Spong, Alan Moore, Helen Taylor, Geoff Lovejoy, Doug Jeatt, Les Copley and Bob Napper. Unseen, in each case was Tony Wright, who fortunately was there to record the scene.

For many years the staff, both past and present have enjoyed a pre-Christmas get-together, either at The Hernes Oak or The Squirrel.

The author also wishes to dedicate this volume to all those who have made *White Bus Services* what they have been over the past 85 years.

157

Reg. No.	Chassis make/model	Bodybuilder and type	Layout	Date new	Date Acq	Date out
MO 330	Ford T 1-ton	Not known	B14R	Jul-22	New	Sep-29
MO 1512	Ford T 1-ton	Not known	B14R	May-23	New	Jul-28
MO 5816	Republic 10F	Not known	B20F	Jul-25	New	Jun-31
RX 2847	Dennis G	Kenning?	B14F	Jul-28	New	Jul-30
PW 9949	Chevrolet LM	Waveney	B14F	Jul-27	Jul-31	Mar-34
FG 3768	Reo Pullman Junior	Not known	B16D	Mar-28	Aug-31	Jul-36
RF 4355	Dennis G	Not known	B18F	Mar-28	Sep-33	Jul-36
JB 4838	Dennis Ace	Abbott	B20F	Sep-34	New	May-43
JB 9468	Dennis Ace	Dennis	B20F	Jul-36	New	Aug-50
JB 568	Commer Centaur T20X	Not known	B20F	Jul-32	Aug-36	Sep-39
BRX 865	Dennis Falcon	Dennis	C26F	Aug-39	New	Aug-50
KX 7454	Dennis Dart	Park Royal (1939 re-body)	C20F	Aug-31	Apr-43	Jul-49
BUN 677	Bedford WTB	Duple Hendonian	C26F	Jul-39	Apr-48	Oct-55
EUX 524	Commer Commando 17A	Harrington	C30F	Dec-48	Jul-50	Sep-55
CMO 495	Bedford OWB	Mulliner	UB32F	Oct-42	Nov-50	Sep-55
ATL 835	Bedford OWB (Perkins)	Duple MkII	B30F	Jan-46	??-51	Mar-57
DRX 296	Bedford OB	Mulliner	B28F	May-47	Mar-55	Sep-56
MVX 508	Bedford OB	Duple Vista	C29F	Feb-48	Jul-56	Nov-61
KXU 673	Leyland Comet CPO1	Windover	C33F	May-50	Aug-56	Jan-60
CMO 773	Bedford OWB	Duple	UB26F	Jun-43	Oct-56	Aug-59
OPG 991	Albion Victor FT39N	Allweather	FC31F	Oct-50	Oct-58	Jan-60
JXH 661	Leyland Tiger PS1/1	Harrington	C33F	Mar-48	Jan-60	Jun-62
NKT 934	Beadle-AEC	Beadle	FC39F	Mar-51	Jun-60	Jul-64
KOE 207	Morris-Commercial OPR	Plaxton/Beccols (r/b 1950)	C22F	Nov-49	early-61	not run
JXH 719	Bedford OB	Duple Vista	C29F	Apr-48	May-61	Jun-63
TMY 950	Bedford OB	Duple Vista	C29F	Jun-50	Nov-61	Feb-66
OPB 750	Albion Victor FT39N	Allweather	FC31F	Feb-50	Oct-62	Apr-65
SAR 128	Bedford SBG	Duple Vega	C36F	Apr-54	Mar-64	Jun-68
LEA 500	Bedford SBG	Burlingham Baby Seagull	C36F	Apr-55	May-65	Mar-68
YFH 53	Ford 570E	Plaxton Consort IV	C41F	Oct-59	Oct-67	Jan-73
CU 9756	Albion Aberdonian MR11N	Weymann	B44F	Oct-57	Jun-68	Aug-71
VXP 508	AEC Reliance 2MU3RV	Weymann Fanfare	C41F	Mar-59	May-70	??/75
NRD 371	Bedford SBG (diesel)	Duple Vega	C41F	Jun-57	Nov-70	Jul-75
6881 R	Bedford SB5	Yeates Pegasus	DP45F	Jun-63	Sep-72	Oct-75
LJB 403L	Bedford YRQ	Duple Viceroy	C45F	Oct-72	New	Sep-79
SNK 255N	Bedford YRQ	Willowbrook Expressway	B45F	Sep-74	New	Apr-98
HRO 958V	Bedford YLQ	Duple Dominant	B45F	Aug-79	New	Oct-O3
STL 725J	Bedford YRQ	Willowbrook OO1	DP43F	Feb-71	Jul-81	Apr-98
EAJ 327V	Bedford YMT	Plaxton Supreme IV	C53F	Aug-79	Apr-85	Feb-O4
JMJ 633V	Bedford YMT	Plaxton Expess IV	C53F	Jan-80	Jan-87	Dec-O7
GNH 530N	Bedford YRQ	Willowbrook OO1	B45F	Oct-74	Mar-87	May-87
GNV 983N	Bedford YRQ	Willowbrook OO1	B45F	Nov-74	Mar-87	Feb-96
RLW 778R	Bedford YMT	Duple Dominant I	C53F	Jan-77	Dec-90	Mar-OO
KDT 281P	Bedford YRQ MkII	Plaxton Panorama Elite III	C45F	Oct-75	Dec-90	Oct-O3
GHD 668N	Bedford YRT	Plaxton Panorama Elite III	C53F	Apr-75	Dec-90	Apr-93
B633 DDW	Bedford YNT	Plaxton Paramount II	C53F	Jul-85	Feb-95	Mar-O8
C668 WRT	Bedford YNT	Duple Dominant	B63F	Jun-86	Oct-95	Dec-O7
ODL 632R	Bedford YMT	Duple Dominant	C53F	Mar-77	Sep-96	Dec-96
B542 OJF	Bedford YNT	Duple Laser	C53F	Mar-85	Mar-97	May-O6
D473 WPM	Iveco 49.10	Robin Hood City	B21F	Nov-86	Jul-97	Apr-OO
JNM 747Y	Bedford YNT	Plaxton Paramount I	C53F	Apr-83	Dec-97	Nov-O6
B30 MSF	Bedford YNT	Duple Laser II	C53F	Apr-85	May-98	Dec-O6
D259 FRW	Bedford YNV	Duple 320	C53F	Apr-87	Oct-99	Jul-O5
E849 AAO	Bedford YNV	Plaxton Paramount III	C55F	Dec-87	Nov-99	Jun-O8
TIW 2795	Bedford YNV	Van Hool Alizee	C53F	May-86	Aug-O1	Apr-O5
R89GNW	DAF SB220	Optare Delta	B49F	Apr-98	Apr-O3	Current
S158 JUA	DAF SB220	Optare Delta	B49F	Sep-98	Apr-O3	Current
L522 MDP	MAN 18.370	Berkhof Excellence 1000L	C53F	Apr-94	Jun-O3	Sep-O8
L715 FPE	MAN 18.370	Berkhof Excellence 1000L	C53F	Apr-94	Jun-O3	Mar-O8

Reg. No.	Chassis make/model	Bodybuilder and type	Layout	Date new	Date Acq	Date out
P131 RWR	DAF SB220	Optare Delta	B49F	Jul-97	Jan-O4	Jun-14
W808 AAY	Iveco Euro Rider 35	Beulas Stergo E	C49FT	Mar-OO	Apr-O5	Sep-O5
L561 ASU	Dennis Javelin	) Wadham Stringer	C70F	Dec-93	Aug-O5	Current
L523 MJB	Dennis Javelin	) Vanguard II	C70F	Jan-94	Sep-O5	Current
R674 OEB	Dennis Javelin	Berkhof Excellence 1000L	C57F	Oct-97	Sep-O5	Jan-13
L606 ASU	Dennis Javelin	W-S Vanguard II	C70F	Jan-94	Oct-O5	Aug-13
R714 KGK	Dennis Javelin	Berkhof Radial	C53F	Apr-98	Sep-O6	Nov-11
L321 XTC	Dennis Javelin	W-S Vanguard II	DP70F	Jan-94	Dec-O6	Aug-O9
YJ07 EGD	Optare Tempo X1260	Optare	B46F	Mar-O7	Dec-O7	Current
YJ57 EHV	Optare Tempo X1260	Optare	B47F	Nov-O7	Dec-O7	Current
V917 TAV	Dennis Javelin	Berkhoff Radial	C57F	Jan-OO	Jan-O8	Nov-11
CB53 MTB	Volvo B7R-63	Plaxton Profile	C53F	Nov-O3	Mar-O8	Current
DB53 MTB	Volvo B7R-63	Plaxton Profile	C53F	Nov-O3	Apr-O8	Current
SN55 DVB	Alexander-Dennis	Alexander Enviro 300	B60F	Sep-O5	Apr-O8	Current
SN55 DVA	Alexander-Dennis	Alexander Enviro 300	B60F	Sep-O5	May-O9	Current
LD04 MCT	Volvo B7R-63	Plaxton Profile	C70F	May-O4	Oct-11	Current
MX09 GZL	Alexander-Dennis	Alexander Enviro 300	B60F	Mar-O9	Oct-11	Current
YN10 ABK	Volvo B7R-63	Plaxton Profile	C57F	Mar-1O	Nov-11	Current
YN10 ABO	Volvo B7R-63	Plaxton Profile	C57F	Mar-1O	Nov-11	Current
YN59 GPE	Volvo B7R	Plaxton	C70F	Oct-10	Jul-13	Current
YN59 GPF	Volvo B7R	Plaxton	C70F	Oct-10	Jul-13	Current
YJ14 BBN	Optare Solo M790SE	Optare	B27F	Mar-14	New	Current
YJ14 BBO	Optare Solo M790SE	Optare	B27F	Mar-14	New	Current
YJ14 BBU	Optare Versa V1170	Optare	B44F	Apr-14	New	Current
YJ14 BBV	Optare Versa V1170	Optare	B44F	Apr-14	New	Current

Former Registration Numbers carried by vehicles listed above

JNM 747Y	was ESU 707 from 2/89 to 1997
E849 AAO	was FSU 739 and E849 AAO previously
TIW 2795	new as C328 EME, re-reg as 1287 RU (for '128 TRU') 5/88, to C710 VPM 11/92, to TIW 12/96
R89 GNW	was new in Ireland as 98-D-61842
L522 MDP	was new as 1598 PH becoming L522 MDP 6/03
L715 FPE	became 5881 PH 9/94, reverted to L715 FPE 6/03
L561 ASU, L523 MJB, L606 ASU and L321 XTC formerley carried military registrations 75 KK 25, 47 KL 48, 47 KL 53 and 75 KK 27, whilst XTC had in between been L860 EPM	
V917 TAV	ran as ESU 320 for a time before being acquired as shown, and was originally G12 ELY
LD04 MCT	was originally YN04 WSU

Standard Body Codes as used in this fleet list

Letters before seating capacity			Letters after seating capacity		
B	Single-deck service bus		F		Front entrance
C	Single-deck coach		R		Rear entrance
DP	Dual-purpose single-deck		D		Dual Entrance
U	Built to Utility Standards		T		Toilet Compartment
F	Fully-fronted bodywork		Continental offside entrances not shown		

VEHICLES ON LOAN

Reg. No.	Chassis make/model	Bodybuilder and type	Layout	Date new	Date Acq	Date out
SK07 DYA	Alexander-Dennis	Alexander Enviro 300	B44F	Apr-O7	Sep-O7	Sep-O7
YJ07 EGU	Optare Tempo X1260	Optare	B43F	Sep-O7	Sep-O7	Sep-O7
SN56 AXM	Alexander-Dennis	Alexander Enviro 300	B60F	Jan-O7	Aug-O7	Aug-O7
The above trio of vehicles were demonstrators on loan for short periods						
YJ57 YCF	Optare Solo M880	Optare	B22F	Jan-O8	May-O8	May-O8
This bus was loaned by the makers whilst YJ07 EGD was undergoing warranty work from 27th to 31st May						
KX57 FKW	Enterprise Bus (Hungary)	Plaxton Primo	B28F	Sep-O7	Feb-14	Mar-14
KX57 OVC	Enterprise Bus (Hungary)	Plaxton Primo	B22F	Nov-O7	Feb-14	Mar-14
These buses were hired from Dawson Rentals pending delivery of new vehicles						

Appendix Two — Fleet List of Winkfield Coaches 1955 - 1990

Reg. No.	Chassis make/model	Bodybuilder and type	Layout	Date new	Date Acq	Date out
EUX 524	Commer Commando 17A	Harrington	C30F	Dec-48	Sep-55	May-57
BUN 677	Bedford WTB	Duple Hendonian	C26F	Jul-39	Oct-55	Jun-56
The above two vehicles were brought out of the White Bus business when Winkfield Coaches was formed						
GOM 658	Commer Commando 17A	Plaxton K3/46	C30F	Oct-46	Oct-55	Jan-60
The above vehicle technically passed to White Bus in January 1960 but was not used						
JXH 719	Bedford OB	Duple Vista	C29F	Apr-48	Jun-56	May-61
The above vehicle was acquired by Winkfield Coaches and later sold to White Bus for further service						
SMU 212	Maudslay Marathon III	Plaxton Crusader	FC35F	Apr-51	May-57	Feb-64
BJP 76	Bedford SBG	Duple Vega	C38F	Apr-55	Jun-59	Jun-79
BDC 387	Bedford OB	Duple Vista	C29F	Nov-50	May-61	May-67
MOW 501	AEC Reliance MU3RA	Burlingham Seagull	C41F	Jun-54	Mar-62	Jun-71
206 VHX	AEC Reliance 2MU3RA	Plaxton Panorama	C41F	May-59	Mar-64	Mar-75
8122 PU	Ford 570E	Duple Yeoman	C41F	Aug-60	Apr-67	Sep-69
YVA 781	AEC Reliance 2MU3RV	Duple Britannia	C41F	Jan-61	Aug-69	Sep-72
BOO 165B	Bedford VAL14	Plaxton VAL	C52F	Feb-64	May-71	May-73
HAR 213C	AEC Relance 2U3RA	Plaxton Panorama II	C40F	May-65	Jun-72	Oct-78
GUR 416L	Bedford YRQ	Duple Viceroy	C45F	Sep-72	New	Mar-79
4402 NX	Bedford SB1	Plaxton Consort IV	C41F	Mar-60	Mar-73	Mar-76
HTF 449F	Bedford VAL70	Duple (N) Viceroy 36	C52F	Apr-68	Mar-73	Oct-74
PLG 702H	Bedford VAL70	Plaxton Panorama Elite	C52F	Mar-70	Sep-74	Dec-76
81 DBL	Bedford SB8	Duple Bella Vega	C41F	Mar-63	Nov-75	Sep-80
TCG 333M	Bedford YRT	Duple Dominant	C53F	Feb-74	Sep-76	Aug-81
KDT 281P	Bedford YRQ	Plaxton Panorama Elite III	C45F	Oct-75	May-79	Dec-90
52 BDL	Bedford VAL14	Plaxton VAL	C52F	Jun-63	Sep-80	??/84
RLW 778R	Bedford YMT	Duple Dominant I	C53F	Jan-77	Jul-81	Dec-90
GHD 668N	Bedford YRT	Plaxton Panorama Elite III	C53F	Apr-75	Aug-83	Dec-90

Notes-

KDT, RLW and GHD passed into White Bus ownership, further details of use under the main fleet list

Appendix Three — Current Routes Operated

Route No.	Route Description			Type and Serving
O1	Windsor - Great Park - Cheapside - Sunninghill -Sunningdale - Ascot			Public Mon. - Sat.
W1	West Windsor circular local service via Clewer and Dedworth			Public Mon. - Fri.
P1	Windsor Park-and-Ride - Home Park - King Edward VII Car Park - Datchet			Public Mon. - Fri.
P2	Windsor Park-and-Ride - Centrica Car Park (Clewer) to Town Centre			BH and specific days
O4	Maidenhead - Bray - Holyport - Moss End - Newell Green - Warfield			LVS schooldays
O6	Chalfont Common - Gerrards Cross - Stoke Poges - Slough - Windsor			LVS schooldays
O7	Sunningdale - Sunninghill - South Ascot - Ascot Station - Heatherwood			LVS schooldays
O9	Crowthorne - Camberley - Lightwater - Sunningdale - Sunninghill - Ascot			LVS schooldays
23/23A/23B	Winkfield - The Village - Royal Lodge - Old Windsor - Windsor			Windsor schooldays
24*	Various points in Winkfield, North and South Ascot (see timetables)			Charters schooldays
28	Winkfield - Cranbourne - Woodside - Cheapside - Sunningdale - S'hill			Charters schooldays
39A/39B	Luckley Road - Barkham Ride - Garrison - California X - St Sebastian's			St. Crispins sch.days
83	Fifield - Holyport - Moneyrow Green - Paley Street - Woodlands Park			Cox Green sch. days
88	Fifield - Oakley Green - West Windsor - Dedworth - Clewer - Windsor			Windsor schooldays

Notes-

Route 01 has the relevant schooldays journeys of Routes 07, 23 and 23B shown on its tables

* Route 24 has variants 24A, 24B, 24C, 24D and 24E, also with a relief vehicle as 24R and the late bus 24L

Routes 39A and 39B have slightly varying stopping points according to assignments

Route 88 calls at points to serve St. Edward's, Windsor Boy's, Windsor Girl's and Trevelyan Schools

For fuller details of routes, times, fares and periods of operation see the White Bus Services website